Kentucky Maverick

KENTUCKY MAVERICK

The Life and Adventures of Colonel George M. Chinn

∽

CARLTON JACKSON

UNIVERSITY PRESS OF KENTUCKY

Scholarly publisher for the Commonwealth,
serving Bellarmine University, Berea College, Centre College of
Kentucky, Eastern Kentucky University, The Filson Historical Society,
Georgetown College, Kentucky Historical Society, Kentucky State
University, Morehead State University, Murray State University,
Northern Kentucky University, Transylvania University, University of
Kentucky, University of Louisville,
and Western Kentucky University.
All rights reserved.

Editorial and Sales Offices: The University Press of Kentucky
663 South Limestone Street, Lexington, Kentucky 40508-4008
www.kentuckypress.com

Photos are from the author's collection.

Library of Congress Cataloging-in-Publication Data

Jackson, Carlton, 1933-2014.
  Kentucky maverick : the life and adventures of Colonel George M.
Chinn / Carlton Jackson.
    pages cm
  Includes bibliographical references and index.
  ISBN 978-0-8131-6105-1 (hardcover : acid-free paper) —
  ISBN 978-0-8131-6107-5 (pdf) — ISBN 978-0-8131-6106-8 (ePub)
  1. Chinn, George Morgan. 2. Kentucky—Biography. 3. Marines—
United States—Biography. 4. United States. Marine Corps—
Officers—Biography. 5. Military art and science—Technological
innovations—United States—History—20th century. 6. Historians—
Kentucky—Biography. 7. Kentucky Historical Society—Biography. 8.
Businessmen—Kentucky—Biography. 9. Bodyguards—Kentucky—
Biography. I. Title.
  F456.C487J33 2015
  976.9'043092—dc23
  [B]   2015017763

This book is printed on acid-free paper meeting
the requirements of the American National Standard
for Permanence in Paper for Printed Library Materials.

Manufactured in the United States of America.

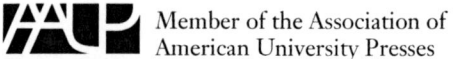
Member of the Association of
American University Presses

This book
is dedicated to the GIs
who fought our wars for us.

Thank you.

Contents

Photographs follow page 96

Introduction

ॐ

A Maverick from a Family of Mavericks

When Colonel George Morgan Chinn took his final separation papers from the U.S. Marine Corps, he wore ribbons and medals earned in four conflicts: World War I, World War II, Korea, and Vietnam. Just how did he manage to take part in all these wars, especially World War I? His birthday was January 15, 1902, which would have made him just fifteen when the United States began to send troops over to France.

The answer has to do with his upbringing and schooling. His earliest "intellectual" memory was of his mother reading the Deer Foot series to him. This series was about pioneer life around Harrodsburg, Kentucky, including the areas of Danville, Lexington, and Nicholasville. "There's more real history in this area, what's called the 'Big Settlement' area, than any place in the United States," he asserted.[1] Always included in the stories were guns, and even as a child, he made the connection between Kentucky history and firearms. Moreover, he was familiar with guns and ammunition early in life; his father, George P. Chinn, was, for some time, the sheriff of Mercer County, a position that put young George in the presence of weaponry.

Later, when his father was warden of the Kentucky penitentiary in Frankfort, George Morgan Chinn frequently accompanied him. Just up the street from the prison was the state arsenal, with Major C. W. Longmire in charge. As it happened, there was

1

a Gatling gun on the premises, and Longmire allowed George to play with it. Before long, George could take it apart, although he had some trouble getting everything back together again. Longmire patiently taught the boy how to reassemble the Gatling gun. From these experiences came young George's interest in—perhaps obsession is a better word—guns of all kinds and descriptions. No wonder he grew up to be the nation's, if not the world's, number one military weapons expert.[2]

George Morgan grew up at Mundy's Landing, on the Kentucky River, never losing his love or fascination for this area. He and his dog, Tauser, were often seen fishing on the river after Touser had "dug him a can of worms or caught a bucket of minnows."[3] "I didn't grow up," he told friends, "I swam up."[4] Consequently, Chinn "swam up on the mighty river of Kentucky history."[5] From Mundy's Landing in Mercer County's Brooklyn community, one could see another place of treasure: the palisades, massive rock formations above the Kentucky River. These palisades had future consequences for George Morgan Chinn.

George Sr. made up his mind early on that his son was going to be a military man, because—as the son joked—family members thought he'd look nice in a uniform. In preparation, his father sent George Morgan to a one-room schoolhouse in the Harrodsburg area called Braxton Hall. Many of the military cohorts he came across later introduced themselves pompously as "Major So-and-So, Cal Tech," or "Captain Somebody, MIT." Invariably, Chinn would introduce himself as "Major Chinn, Braxton Tech." He later remarked, "I haven't had anybody yet ask me where in the pluperfect so-and-so is Braxton Tech. No one ever questioned it [the title] and they always gave a knowing glance like they knew it well."[6]

The students at Braxton "Tech" faced a battery of questionable teaching techniques. One "old maid" teacher put them through numerous classroom spelling bees. If a student missed just one word, he or she (girls were on one side of the one-room school and boys on the other) would have to go to the end of the class as it lined up to show off its spelling abilities. If, heaven forbid, a student missed *two* words in a row, he or she not only had to go to the end of the line but recite, in correct order, the name of every president from George Washington to Woodrow Wilson. George became so efficient here that it took him only a few seconds to declaim, "Washington, Adams, Jefferson . . . Roosevelt, Taft, Wilson!" One gets an insight into his behavior when he admits that he could go backward from Wilson to Washington in short order as well.

He did, however, learn one substantial thing in school about engineering: the shortest distance between two points is a straight line. He argued that engineers, particularly those with college degrees, complicated things by lessening obstructions, whereas Chinn wanted to *add* things. To *add* parts of a machine gun to make it work is, he said, simplicity rather than complexity. "Simplicity is the highest form of art," he liked to say.[7]

His tenure at Braxton "Tech" helped to forego regular high school and got him into the Millersburg Military Institute, a few miles up the road from Lexington. He did not particularly like this early military experience. He called the Reserve Officer Training Corps (ROTC) the "Saturday Afternoon Tea Club."[8] Rising to captain of Company C, he excelled at track (despite his tendency toward overweight) at Millersburg and set a state record in javelin throwing that stood for many years.[9]

He left the campus unauthorized on numerous occasions, for which he received whippings administered by Colonel C. M. Best, superintendent of the school. Chinn described himself as a "hard-headed boy"—after Superintendent Best's sixth beating, he exclaimed that he would not yield to any more assaults on his body. In the past, any "lip" from a student, any insubordination, as that word was defined by higher-ups, brought on severe discipline and ten-hour shifts of guard duties, during which there was little or no rest. Nevertheless, at Millersburg Chinn learned the discipline and determination that served him so well in later years. When he came out of the academy, he could, he said, "out walk anybody on Earth."[10]

He was fourteen years old when he entered Millersburg Military Institute and nineteen when he was graduated (1920) as a cadet major in a class of nine. Each member of his class at Millersburg received a Victory Medal, "although he was never in theater."[11] "I resented missing World War I," Chinn asserted, but actually, in a sense, "I did not miss it." The cadets at Millersburg, including George Morgan Chinn, were scheduled to be commissioned as officers in December 1918, and transferred immediately to France. The war ended with the Armistice on November 11, 1918, just one month short of the Millersburg graduates' deployment.[12]

George Morgan Chinn was a ninth-generation Chinn in what became known as the Bluegrass region of Kentucky. It was bordered by mountains on the east and south, the Ohio River to the north, and on the west by a "treeless wasteland called the 'Barrens.'"[13] The first settlers coming into the "Great Settlement area," Chinn asserted, were veterans of the French and Indian War (1756–1763), not, as commonly supposed, the American Revolution. In fact, he claimed, "The revolutionary soldier . . . was a 'Johnny-Come-Lately'" to the area. Since Harrodsburg was settled in 1774 and since "you didn't have the Revolution until '76," it stood to reason, at least for Chinn, that French and Indian veterans predated those from the Revolution.[14] (This statement, and many others Chinn would make in the years ahead, infuriated "professional" historians.) He also argued that few if any Indians were in the area when the white settlers came. "No white man," he said, had "died of arrows" when Harrodsburg was founded in 1774. The first pioneers were from Pennsylvania; later depictions of immigrants coming to the area by way of the Cumberland Gap were, like their revolutionary counterparts, "Johnny-Come-Latelys."[15] This and other provoking statements led Kentucky's leading historian, Dr. Thomas D. Clark, with whom Chinn was frequently at loggerheads, to pronounce that Chinn was "very much devoted to Mercer County," seeing it as the "center of the universe."[16]

But in fact Mercer County could very well have been seen as the center of the universe by the entire Chinn family. The first of them came in 1781, "and we've moved three times since then, no more than six miles."[17] Rawleigh Chinn migrated from Virginia to "The Settlement" and claimed some land, but moved back east to serve with George Washington's Continental army. After the Revolution, he returned to the Bluegrass, where the Chinn family has remained ever since.[18] Some of the Chinns developed a near-proprietary attitude toward the region, which included "protecting" its history from those whose interpretations might differ from their own.

If George Morgan Chinn was an extraordinary individual, he came by it honestly. The Chinn family abounded in colorful, often controversial, characters. When in 1906, his father became warden of the Kentucky penitentiary in Frankfort, he abolished the practice of whipping prisoners, especially for minor offenses. Chinn Sr. "believed there was good in everybody," and lost his job as a result of that philosophy; the "authorities got rid of him."[19] For several months after his dismissal, Chinn Sr. talked and wrote about the disreputable conditions he had found in the Frankfort penitentiary when he first took over as warden. Calling for an investigation, he spoke of prisoners being "gagged and tied to a board and then whipped with a bloody strap." Some prisoners were so desperate they cut off their fingers to avoid punishments for not doing tasks assigned to them or for not being physically able to perform them.[20] Dr. Joseph Barr, a physician at the penitentiary, said that conditions at the facility were so poor that "75 per cent of the deaths that have occurred at the prison during the last two years had been from tuberculosis," concluding, "This penitentiary has become a veritable death trap."[21] This shameful operation was run by private corporations, primarily the Hoge-Montgomery Company, in association with the three-man prison commission.[22]

And then John Pendleton Chinn, Colonel Jack, George P. Chinn's father and George Morgan Chinn's grandfather, joined the fray. From his position in the Kentucky legislature, Jack intended to "unmask the grasping shylocks and grafters, that are coining the blood of convicts into money for their own pockets."[23] For several weeks Jack and his son, George, were in the news. A reporter once told Jack Chinn that Chinn had been interviewed so often there seemed to be nothing new to write about him. Jack Chinn replied that he didn't mind what the reporter wrote . . . so long as Chinn was not ignored. That was an attitude passed down to Jack's grandson, George Morgan Chinn. He'd rather be told to "go to hell" than be ignored.

John Pendleton Chinn helped considerably to build the family aura. He became well known not only throughout Harrodsburg and the Great Settlement area but in the state and nationally. Opinions of him differed widely; he was regarded as a great ambassador for the Bluegrass State by some and as a brawler and general ne'er-do-well by others.

He was, first and foremost, into horse racing, known widely as Kentucky's most famous turf man. After several mixed seasons, Jack Chinn and his partner, George Martin, bought a thoroughbred named Leonatus (1880–1900). His sire was Longfellow and his dam Semper Felix. Leonatus was a 2 to 1 favorite to win the Kentucky Derby of May 23, 1883. Jockey William "Billy" Donohue fought off competitors such as Drake Carter and Lord Raglon to win the classic horse race. In fact, Leonatus finished three lengths ahead of Drake Carter, his nearest competitor. Leonatus's owners collected $3,760 for this Derby run. Needless to say, both Chinn and Martin were overwhelmed with joy. Leonatus also benefited: when he was led into the Winner's Circle, apparently famished after such a strenuous workout, he proceeded to eat the presentation roses.[24]

There were other horses and races ahead for Jack Chinn. Harrodsburg, for example, was an entry in the Kentucky Derby in 1886, but lost out to Ben Ali. Jack was also affiliated with a horse, George Smith, who won the Derby in 1916. George Smith had been bred by Fred Forsyth and Jack Chinn (some sources say it was Jack's brother Christopher Chinn, known as Kit, who was the partner), foaled at the Mountain Blue Farm in Harrodsburg. But the drama and excitement of the Leonatus victory never quite returned. Instead, Jack devoted much of his time to the creation of laws, rules, and regulations for racing, and for building new racetracks.

For instance, he gave much attention to a local racetrack known as Crab Orchard, just outside Harrodsburg. William Whitley owned the land on which an oval racetrack (thought to be the first of its kind in America) was built in 1778, nearly a century before the first Kentucky Derby was run in Louisville. Races were held at Crab Orchard yearly, attracting visitors and bettors from all over Kentucky, the region, and even the nation.[25] The surroundings helped to bring in people for the racing events: four mineral springs graced the area, and one freshwater spring was more than suitable for making fine and tasty whiskey, which "brought in the elite from all over the South."[26]

Whitley was a fan neither of the British nor of British racing. They ran their horses clockwise on grass at their famous racing establishments at Epsom Downs and Ascot; Whitley directed his horses counterclockwise on clay, a pattern that supposedly set the anti-British style for horse racing throughout the United States.[27] (Today, Kentucky Downs in Franklin, Kentucky, claims to be the "one of a kind" European-style turf course in America; if so, this is the only racetrack in the country that countermands Whitley's plans.)[28]

Jack Chinn had been a Confederate officer during the Civil War, and he believed his "southern" propensities helped to divert attention from his favorite racing venue, Crab Orchard, to Louisville. His grandson, George Morgan Chinn, went even further; he asserted that if the South had won the War between the States, the Kentucky Derby would have been held at Crab Orchard instead of Louisville.[29] Again, this was the kind of statement that created much controversy among horsemen; the debate goes on to this very day. George Morgan, Kit, and Jack—all three of them delighted in stirring up debates and arguments.

Jack and Kit Chinn fretted throughout the 1890s that certain unsavory practices had crept into racing that must be stopped. In 1886, for example, a private bookmaker had "cornered" the market on "pooling privileges," or wagering.[30] This practice cut out many "public" bookmakers and actually caused a decline in attendance. Jack was instrumental in getting the Kentucky Racing Commission enacted into law by the state legislature. Its actions, Chinn hoped, would regulate the racing industry in fair-minded ways, setting rules and standards to be met by all the "jockey" clubs in the Commonwealth.

The organization meeting of the commission took place on April 18, 1906, in Frankfort, Kentucky. John Pendleton Chinn was named its chairman. At a subsequent meeting on April 23, 1906, at Lexington, the commission created rules of conduct for board members, jockeys, stewards, owners, and trainers: they were to be polite in all circumstances and never swear in public. In short, the commission became the all-embracing authority of Kentucky racing.

Several clubs and individuals went before the commission when it met in Lexington on April 23, seeking licenses for racing dates and looking for guidance on procedures to be followed during the races themselves. Mr. J. Winn, manager of the New Louisville Jockey Club, representing riders of racehorses in

Louisville, asked for racing dates of May 2–May 29, a request that was immediately granted. Representatives from the Latonia Jockey Club in Covington were also present. They wanted to start their racing on May 30, just one day after the Louisville meet ended. This was the very thing the commission desired: racing dates that did not conflict with one another, assuring large attendances at each of Kentucky's major racetracks.

Besides immersion in all aspects of Kentucky's horse racing, Colonel John P. Chinn was also interested, at least for a time, in regional and state politics. He ran for Congress once, for Kentucky's Eighth District, but for reasons never disclosed, dropped out of the race even before the primary.[31]

He did serve numerous terms in the Kentucky General Assembly, garnering a great deal of influence over candidates and races. He appeared to be a "kingmaker" instead of the king himself. Widely known as a strong advocate for free silver (as opposed to gold as the chief standard of currency), he relished visits to political meetings, especially those friendly to silver, where he would hold his gold-plated walking cane up high, so everybody could see it.[32] Whether his intention was to make a statement, show off just to be the center of attention, or make fun of the gold supporters led to much speculation in political circles. His action was regarded as "arrogance" by some, "independence of thought" by others. Whatever, it was a widely talked about act of showmanship that entertained just about everyone in the audience.

Louisville Times columnist Tandy Ellis reported that in the 1890s Jack Chinn supported silver candidate J. C. S. Blackburn, always complaining that the latter was not forceful enough against the "sound money" gang or the "gold bugs," as they were commonly called. Chinn told Blackburn to "crack down on them." Blackburn finally got around to discussing and attacking Kentucky's gold advocates, even within the Democratic Party itself. He orated: "Their [gold Democrats'] souls are blacker than the walls of hell. The vilest criminal who treads the molten path of that inferno would blush if found associating with them, and I, for one, wish and hope that some vast crater would open up and swallow them, each and all of them, down to depths abysmal, where their sin-seared souls may burn eternally."[33]

Blackburn figured that this fiery speech would deeply impress his friend Jack Chinn. It did, but from Blackburn's point of view the wrong way. Despite all Blackburn's hyperbole Chinn told him, "Fire Up, Joe! You can't win this fight that way. You are too damn cold and intellectual!" In addition to Blackburn, many other politicians embraced this advice (though Chinn never did tell them how to make their speeches less cold and intellectual) from the colonel and went on in later life to be quite happy about it. Colonel Chinn apparently loved to lead and influence people to embrace his ways of thinking (a characteristic that his grandson, George Morgan Chinn, also possessed). This was true with horsemen, politicians, and family members.

George Morgan Chinn grew up in his grandfather John Pendleton Chinn's household. His mother, Anna, died when he was eight; after her death, George and his father lived with Grandfather John and Grandmother Ruth Grace Morgan (hence George's middle name) at the Chinn family farm in Mercer County and then at Mundy's Landing.[34] One of George's father's cousins, Hannah Morgan, who had been in Utah teaching school to various tribes of Indians, came back to Harrodsburg to help raise George Morgan and to care for the grandmother, overtaken by age. George often remarked that Grandfather Jack was one of the most "even-tempered" persons he had ever seen—he was "mad all the time."[35]

Young George loved hearing stories about John's racing accomplishments, his short-lived political career, and the statewide influences he had amassed through his various activities. Invariably, telling these stories, Grandfather almost always got around to telling the tale of William Goebel, an elected governor who, on January 30, 1900, was slain on the steps of the state capitol building in Frankfort.[36]

John Pendleton Chinn was one of two bodyguards protecting the new governor on that fateful day in the state's capital. He was standing less than two feet away from Goebel when the assassin fired, and one of the bullets went through his shirt collar. Chinn reached out to the stricken governor and unsuccessfully tried to keep him from falling. "He was dead weight as he hit the ground," John asserted.[37] An enterprising (or perhaps "foolhardy" is a better word) newspaper man, Theodore Hallam, claimed that John Chinn himself had shot the governor. After arguing the matter for a while, the two, Hallam and Chinn, fought it out. Chinn won, and Hallam admitted that he had concocted the whole story for publicity reasons.[38]

Over the years, young George M. Chinn heard many details of the event. "I often heard the Goebel story," Chinn related, "especially when Grandpa had a few drinks." "As the jug of whiskey got lower," Chinn remembered of his grandfather, "his voice got louder." In later years, George liked to joke (at least, hopefully he was joking) that "the only mystery" about the Goebel era "is how the hell did that Yankee from Pennsylvania ever get elected in Kentucky?"[39] This was not the last "dig" that George Morgan Chinn would ever make about Pennsylvania; this was especially true when the subject of "Kentucky" versus "Pennsylvania" rifles came up. The two states carried out a war of words (mostly in a friendly manner) about which was first to invent this weapon, which was at that time quite revolutionary.

Symbolic of the low regard in which some Kentuckians held George Goebel, George Morgan Chinn also loved to describe to his friends and colleagues how Grandfather John and many of his "drinking" compadres got together each Sunday afternoon in the Chinn backyard to "plink" Governor Goebel. "Plinking" involved firing at a nonliving entity, whether it was a leaf on a tree, a bud on a plant, or a figure of a human being drawn or etched on a piece of cardboard. For the most part, it was a harmless display of marksmanship, but each time either John or one of his guests scored a point against "Goebel," a great roar of approval went up into the air, heard literally for miles around.

Jack P. Chinn died in 1920. Jim Miller, columnist for the *Harrodsburg Herald*, gave what was probably the most accurate obituary. John Pendleton Chinn "was said to be by birth a Kentuckian, by profession a turfman, by association a pal and by interest a chronic scrapper."[40] Jack Chinn would have agreed with these words just as much as his descendant George Morgan Chinn did: he savored every word of the obit.

Another of George Morgan's independently minded relatives was his great-uncle Christopher Chinn (b. 1846), who certainly contributed his share of creating and maintaining family mystiques. Chris, known widely throughout the settlement as Kit, owned a slave named Madison. For many years Madison had been a bell ringer and caretaker for the local Methodist church. Each time a congregant died, Madison rang the bell once for each year of the deceased's life.[41]

After many years of servitude to Kit, Madison himself died in 1860. Kit ordered that funeral services for Madison be conducted at the all-white Methodist church, much to the opposition, in some cases expressed violently, of the church members. Nevertheless, the bells tolled for Madison, and the white minister intoned the last rituals.

After the funeral, Madison's body was taken to the tiny African Methodist Church on Harrodsburg's Broadway Street. It was the practice at the time for "higher-society individuals" to send out black-bordered funeral notices, and this Kit did for Madison: "The Servants of Your Family are Requested to Attend the Funeral Of Madison, the servant of Judge Chinn, this evening [September 13, 1860] at 3 o'clock at the African-American Church. From thence to Spring Hill Cemetery. Funeral services by the Rev. Mr. Gould."

As though the uproar in Harrodsburg and surrounding areas was not bad enough over having Madison eulogized in a white church, Spring Hill Cemetery, where Kit proposed to bury Madison, was designated whites only. It was on a piece of land that had once been owned by Governor Beriah Magoffin; white people in the community had always been welcome to buy into it. Kit did so, and he wanted to bury Madison in the Chinn family plot. The public outcry increased considerably when Kit's plans became generally known.

But there was nothing the protestors could do about it. Investigators found no laws, local or otherwise, that ruled against burying a black in a white cemetery. Madison was, therefore, the first "of his race to rest in Spring Hill." And, it turned out, the last one. The Harrodsburg Town Council lost no time, with the support of the local populace, in passing an ordinance against the mixing of races in community-owned cemeteries.[42]

The story of Madison and Christopher Kit Chinn was told and retold by the citizenry for the rest of the nineteenth century and beyond. Kit's action helped to put an antislavery face on the Chinn family, one that supported the idea of freedom and equality as being essential elements of democracy. George Morgan asserted that his was a family of "rugged individualists." He added that "if you are a genealogist you hate relatives who went to church every Sunday and paid their taxes and never ran off with their neighbors' wives. You can't find out anything about them. But I'll just tell you mine," he went on enigmatically, "were easy to find."[43]

The Chinns had an interest in marble, of which good supplies were found on the Chinn property, along with calcite, adding to the wealth and illustrious standing of the Chinn family both locally and regionally. They never let their good fortunes "go to their heads." Happy and profitable days accrued for the family with marble and calcite, especially the latter, for it was widely used by manufacturers all over the country, most notably

Corning Glass of New York, which, among other things, made telescope lenses.[44]

As George Morgan's graduation from Millersburg Military Institute loomed, he was occupied with thoughts of what he would do next. He was nineteen and ready to move on with his life. He applied to Centre College, just down the road from his home in Harrodsburg, and was, to his pleasure, accepted. He majored, he said later, in "football and penmanship."[45] Actually, physical education was his major, and this status helped to get him on the football team, a team that made U.S. football history in October 1921. That was when the Centre Praying Colonels defeated the mighty Crimsons of Harvard University. Colonel Chinn marveled at this great accomplishment, of which he had proudly been a part, for the rest of his life.

1

∾

What's in a Name?

Over the years, all members of the Chinn family of Mercer County, Kentucky—George Morgan Chinn and his forebears—endured their share of teasing, being called "Chinn Ups," "Chinny-Chin-Chinns," and (in the colonel's case, because of his weight) "Double-Chinn."[1]

The original family name was des Chynn, of French Huguenot derivation. When the family arrived in Mercer County, they immediately became aware of the ill will in the area between the English and various Indian tribes. With the latter receiving strong support from the French, war loomed many times between these two adversaries. "It was better to change the name [of the family] when moving to an English settlement. It was a lot quicker," the colonel frequently said, "than trying to explain that a French Huguenot was a Protestant refugee."[2] The des Chynns, then, changed their name in the eighteenth century for the same reason Germans in the United States changed theirs during World War I: to protect themselves from the prejudice and hostility of other nationalities.[3]

Years later, when Chinn was in the Orient, he remarked, "'Chinn' must be 'Smith' in Chinese, there were so many of them. Turn a corner in Korea, there's a Chinn; go down a new street in the U.S., there's a Smith."[4] He joked about owning an interest in a Chinese laundry. While in Vietnam George discovered that *le chien* in French-speaking Vietnam meant "dog." Many times adults laughed when introduced to Lieutenant

Colonel Chinn, while children frequently shied away from him, believing themselves to be in imminent danger from an ill-tempered canine. Military writer George Kontis's first impression of Chinn was that "he must be Chinese." A "big guy," he "looked like a retired Sumo wrestler."[5]

The des Chynns were not the only family to change their name when they came to the Big Settlement area. Many others of French background, or even with names that sounded French, did likewise. Also, some families' names were changed simply because of clerks mishearing the name and writing it down wrong. Chinn illustrated this point with two examples. When one family left North Carolina, Chinn said, its name was Ishamerial. In time, this name was written by a clerk or clerks as Ishmeal. Then, finally, as Chinn describes it, the family came to Harrodsburg, by which time the name had been changed to Isham. With perhaps the unintended help of county clerks, the family name ultimately became Isom.[6] From Ishamerial in North Carolina to Isom in Kentucky was a name-changing path of the kind many settlers experienced.

If the citizens of a community found someone who could read and write, they made him a clerk. In due time, the clerk became the most important man in the village, in fact, a "tsar." If he miswrote the name of a land claimant, that misspelling continued well into the future, perhaps forever. If a contemporary claimant did not use the name that appeared on claims, or was not able to document how his name had changed over the years, he would not get the land. A prime example of such a name change was the case of Abraham Linchorn, who happened to be President Abraham Lincoln's grandfather. *Linchorn* was the way the county clerk heard the name, and that's the way he wrote it. Even in modern times, if you want to inherit that property, "you'd better be a Linchorn" as well as a Lincoln.[7]

Another way for a family name to change, or even be discontinued, was to be "daughtered out." Although families "never die," George asserted, on rare occasions, a family has only daughters and when they marry they do, of course, take on the names of their husbands. George claimed that for ninety-two years, the Chinn family had sired only sons, ensuring that the family name would continue in perpetuity. Yet George M. Chinn and his wife, Haldon, raised only one daughter, Ann (Anna). In time Anna married a man named Howells, and they had three children: Ann Howells, Ruth Howells, and Howard (Buddy) Howells II. Thus, from George Morgan Chinn's perspective as maternal grandfather, this particular faction of the family would not carry on the family name.[8]

George came from a long line of Methodists but became a Presbyterian when he married Haldon. Asked by a reporter to what religious denomination he belonged, Chinn answered: "Presbyterian, by marriage."[9] Chinn's wife, Haldon, assumed a humorous nickname that became known far and wide in Harrodsburg and surrounding areas. She was "Cotton" Chinn. She loved it.

In late October 1921, a new "symbol" was added to the periodic table, its origin confusing for just a little while even to teachers of chemistry around the state. It was finally traced to Danville, Kentucky, home of Centre College, undoubtedly the best institution of higher education in the state. The "symbol" was: "C6-H0" and it meant "Centre 6, Harvard 0," except that most Kentuckians wanted to spell it out as ZERO. "C6-H0" was painted on the sides of just about every building in Danville (and even Harrodsburg, a few miles away). Some cattle out in local pastures wore the symbol on their sides, showing one and all that even the cows liked what Centre College had accomplished over the weekend of October 29, 1921.[10] Unbelievably, a young boy rode an "old dairy cow" festooned with the symbol up and down Danville's Main Street.[11] The crowds loved it, cheering on the young lad for even further celebrations of the day.

The object of all this love and adoration was the football game—"the game of the century"—some said, that took place on Saturday, October 29, 1921, in Cambridge, Massachusetts. That was the day the Praying Colonels of tiny Centre College in Danville, Kentucky, kept the mighty Harvard Crimson of Cambridge scoreless, winning the game 6–0. They pulled off this feat in front of between forty-five thousand and fifty thousand fans in Harvard's stadium, Soldiers Field, a number at a football game that no one had ever seen in Danville—or the entire state of Kentucky, for that matter.

The Centre players had a rough field on which to practice, and for certain, this increased their stamina, an advantage when they played on regular fields. It was an abandoned brickyard, and the bricks frequently worked their way to the surface, leading to multiple broken toes and even noses. The Praying Colonels were used to "roughness" on their practice field; undoubtedly, this "toughening process" helped considerably to bring about the "Formula" C6-H0. Also, many claimed that the "rarefied atmosphere" of Kentucky's Bluegrass region, with its moderate climate, helped the Praying Colonels win so much success on the football field.[12]

Why did little Centre (Southern Intercollegiate Athletic Conference) in the middle of Kentucky's Bluegrass region play mighty Crimson Harvard (Ivy League) in the first place? It was not at all unusual for two such teams to be scheduled against each other. In 1920–1921 Centre's rivals consisted of Auburn University (Alabama); Georgetown College (Kentucky); Clemson University (South Carolina); and Washington & Lee University (Virginia). During this same general period, Harvard's opponents were, among others, Princeton University (New Jersey); Brown University (Rhode Island); Penn State (Pennsylvania); Middlebury College (Connecticut); and the University of Georgia.[13]

George Morgan Chinn was a happy participant in all this

football activity. After his graduation from Millersburg Military Institute, his grades were good enough to get him into Centre College in Danville where, as mentioned, he majored in "penmanship and football."[14] In the former category he accomplished a little more than he let on: he wrote a column for the Danville newspaper the *Advocate* called the "Sage of Mundy's Landing." He also wrote a column, "Now and Then," for the Centre newspaper. Editor Enos Swaine said Chinn "could rise to any degree of great intelligence; he was a leader in any organization."[15] Weighing 250 pounds at the tender age of eighteen, he went out for football and easily made the team.[16] Playing as a lineman, he started games in the 1920 and most of the 1921 seasons, but seemed to suffer from a tricky elbow and, more seriously, a separated shoulder caused by an injury in a previous game.[17]

In October 1921, both Harvard and Centre remained unbeaten in the 1920–1921 season. A year or so before, the two teams had squared off against each other, to Centre's disadvantage. The Praying Colonels swore on that occasion that the next time they faced this powerhouse from New England, they'd show them what a little college in Kentucky could do.

Under these circumstances, naturally, the students wanted the administration to cancel classes and allow them to take a special train for 125 passengers from Lexington, a few miles away, to Boston; Danville resident and Centre booster George Joplin had put together a "Harvard Special" to take any Centre student or Danville citizen to the big game. The plan seemed set until the Centre faculty stepped into the fray: excusing students to attend an athletic contest violated the rules and regulations of the Intercollegiate Association and could not therefore be countenanced, even on this momentous occasion. Student councils, prominent Danville citizens, and even the local newspapers excoriated the faculty for this "selfish" and "unreasonable" stance on this extremely important matter. Even the Danville Chamber of Commerce tried to use its power and influence to get the trustees to rescind the faculty ban. All to no avail: many Centre students stayed home, blaming the faculty for their "misfortune." The *Danville Advocate* roared that "the college should not be 'crabbed,' the spirit of the student body should not be dampened, nor should the pride of Danville be humbled in such a manner."[18]

Clear and simple, many Centre professors believed that athletics were becoming too significant at the college, not only to the students but to the public as well, a tendency that had to be discouraged or even repelled. (Chinn had always disparaged his professors at Centre: "Of course, all of the professors are . . . nice about me making up my back work. They are as amiable as Simon Legree was to Uncle Tom.")[19] Despite the faculty uproar, some four hundred students and townspeople did catch a special train from Southern Railroad, though it was not now called the Harvard Special, to attend the game. The students cut classes (and many later suffered the consequences) and the citizens skipped work to attend this event, which even today is still called "The Game of the Century."[20]

Well before the train's 8:00 a.m. departure from Danville on Wednesday, October 26, word had spread uptrack about the impending arrival of the Praying Colonels. At Lexington, not far from Danville, students—including many football players—from the University of Kentucky and also from Transylvania University, turned out to wish the Centre gridironers well in their upcoming confrontation with Harvard. From Lexington, the train traveled on up to Cincinnati, where the passengers were once again met by long lines of well-wishers on the platform. The same was true with Columbus, Ohio, and on up into Pennsylvania lake country, and then New York, Connecticut, and finally Boston!

Chinn reported that head coach Charles (Uncle Charlie) Moran imposed a fairly strict discipline upon both the Centre team and the Danville residents making the trip. He even brought with him the belt he used at practice sessions when a player, in his opinion, did not try hard enough or, worse yet, uttered swear words.[21] Centre's star player, Alvin Nugent (Bo) McMillin's "cussing" on the field extended to words like "Wonderful" and "Marvelous" when he wanted to execute certain plays or admonish players who did not live up to expectations.[22] Many years later, George Morgan told a television reporter that on rare occasions members of the team got into fights—shoving matches and even fisticuffs—but Moran shrugged off such events, saying that "a team that will not fight among itself, won't fight an opponent." He joked with his team members (at least they hoped he was joking) that "if a player died during a game, he'd just pull him off the field, notify the family, and get on with the game" as quickly as possible.[23]

The coach, his assistants, and the entire team prayed before each game. They did not pray for victory so much as that the Almighty would preserve the health and stamina of all the players, keeping them safe from injuries, including those on the opposite team. This was the kind of solicitation that impressed one team after another, and it did have a significant impact on Harvard, both the team and the spectators. In many significant ways, the New Englanders who came out to see the game were so taken by this team from "little" Centre that they rooted for the Kentuckians to win.

On board the train to Boston was the band, the Centre Five, made up of saxophone players, banjos, and percussion; marching through one car after another, the players "knocked off some lively tunes" on the way up east.[24] Riding this train as well was crowd favorite Roscoe Arbuckle Conklin Breckinridge, "Centre's African-American masseur and water carrier." At halftime, Breckinridge "dressed up" and cakewalked. At the 1921 event, for his New England hosts, he "brought down the house," wearing his tall silk hat and yellow vest, "looking like a Kentucky tobacco ad."[25]

At Beantown the Harvard coach and several colleagues met the Centre team and Danville rooters. Chinn related later that one of the Centre players walked up to the Harvard coach and asked, loudly enough for all to hear, "Where's you boys' schoolhouse at?"[26] Not to be outdone in "hillybilly-ese," Chinn chimed in with his own statement: gazing "open-mouthed" at the large crowds, he said, "Gosh, I ain't seed so many people since the calico sale back home . . . in Kaintucky." This behavior caused mirth and amusement among the people who had come to watch the game. They knew the players were joking and accepted their antics in the spirit of camaraderie in which they were intended. They also knew that all the sports pundits in the country gave a 3–1 edge to Harvard.[27]

One person in the crowd made it a point to look up Centre player George Morgan Chinn. A Harvard law student described by friends and acquaintances as always singing, whistling, and sometimes even dancing, his name was A. B. Chandler, but everyone knew him as Happy. Chinn's association with Happy Chandler in the years ahead led to some interesting and significant twists and turns in his life. Now Chandler, formerly a student at Transylvania University in Lexington, escorted Chinn and some other players to the numerous historical sites in Boston.

Then, on Saturday, October 29, 1921, the moment arrived: everyone—players, coaches, students from Harvard and Centre, and spectators from Boston and Danville—converged on Soldiers Field. The *big game* was about to begin. Centre won the toss and Harvard kicked. The first half was described as "rather dull," with constant "plunges through the line" by both teams. Centre showed, however, as one writer presciently said, that "its defense was . . . stronger . . . than last year."[28]

At halftime, Breckinridge's antics and the music of the Centre Five charmed the New England audience. They played "My Old Kentucky Home" before going into a rousing rendition of "Dixie." When the Harvard band took the field, it played an equally gusty version of "Yankee Doodle Dandy." Altogether, at least during the first half, it was a perfect football game: an ideal, crisp fall day and the score 0–0, with the "underdog," Centre, at least holding its own. During halftime, the Centre players huddled while still out on the field, so anxious were they to get back into the game.[29] George Morgan Chinn wore his Centre uniform, but all he could do during the game was stand on the sidelines and chafe because Coach Moran would not let him play because of his injuries.

Early in the third quarter, after Centre had kicked off and then regained the ball, the team from Danville began to feel victory in its grasp. McMillin ran to the right with "Red" Roberts running interference. McMillin saw an opening and dashed to his left. Harvard players Gherke and Johnson stood between McMillin and the goal; when they tried to tackle him, McMillin stopped and twisted around, throwing the tacklers off balance and allowing McMillin to cross the Harvard line for a touchdown. The crowds went wild. Even some of the New Englanders cheered when tiny Centre forged six points ahead of mighty Harvard. Centre kicker Chase Bartlett tried but missed the extra goal. The score was 6–0, Centre.

The Harvard coaches and players believed they could make up the difference because the game still had most of the third and all of the fourth quarter to play.[30] And in fact the Crimson did get within reach on several occasions, carrying the ball to Centre's thirty-two-, thirteen-, eight-, and even to its three-yard line with three downs to go for a touchdown, but the Centre defense held tight on all these occasions.[31] When the final whistle blew, Centre had "humbled" powerhouse Harvard 6–0.

At the Harvard Stadium celebrations after the game, thousands of fans "engulfed" the Centre team and carried Bo McMillin off the field on their shoulders. Huge crowds lined the streets of Cambridge and waved to the Centre players as they made their way back to their hotel.[32] Early the next morning, Sunday,

October 30, the team and its followers boarded a train for Danville, where the exploits of the day before were well known. For the Praying Colonels, however, it was humility that characterized their moods. At an early-morning service, Bo McMillin, the great hero of the day, gathered his teammates around him and prayed: "Oh God, may we be humble in the light of this great victory. Without faith it could not have been won. May we all carry that faith into the outside world, which some of us are soon to face, so that we may be better citizens and a credit to Old Centre!"[33] Interestingly enough, the team rode the remainder of the trip in moods of quietness and contemplation. Were the players reflecting on McMillin's prayer or anticipating that "outside world" Bo had mentioned? As it turned out, it was a bit of both.

The outside world was definitely awaiting them when they arrived in Danville, where most of the six thousand citizens awaited the conquerors. Whistles tooted as jovial fans tried to find a place at the station that offered the best views. A fire truck met the Colonels, all of whom hopped on board. George M. Chinn stood precariously on a fender and held on tightly to one of the truck's hoses. The governor of Kentucky, Edwin P. Morrow, intoned, with tears of happiness rolling down his cheeks, "I would rather be Bo McMillin than the Governor of Kentucky." The town's schools were closed in honor of the returning Colonels and even classes at Centre on Monday, October 31, were canceled.[34]

Even before the Centre team had left Boston, there were rumors among the most die-hard Harvard fans that in the game, Harvard had "kept something back," a canard hotly disputed by both coaches and players of the New England team. The rumors said also that when Harvard next played its "real" rivals (that is, schools like Yale, Princeton, and Penn State) it would pull out all the stops. "These reports have no foundation," Harvard officials announced. "Harvard has done its best in every one of the games on the schedule," and this certainly included the match with Centre College.

Sports headlines across the nation kept repeating the term *upset* in describing the game. And this word more than "upset" a lot of Centre folk, extending from the student body to the administration and townspeople. The feeling on campus was that "no matter how many teams" we defeat; "it was always Harvard they wanted."[35] The team had bowed to Harvard in 1920, 31–14, succumbing again in 1922, 24–10. The big year was to be 1921. And the team and the community wanted the recognition the Praying Colonels so rightly deserved.

The team had begun training for the spectacular win against Harvard in the fall of the previous year. Red Roberts, who had a job at the Danville freight yard, had even asked for the hardest physical work in the place in order to harden his muscles and keep them in fit condition. Robert Myers, athletic chief at Centre (a position comparable to "athletic director," or AD, at most colleges and universities today), told his "boys" that he would "smoke a cigar under a gasoline shower to see you beat Harvard."[36] These words were uttered a year before the famous game at Harvard Stadium. This was no "upset," Centre argued, especially after twelve months of continual muscle flexing through as many types of training exercises as the coaches could imagine. The planning had been too meticulous and thorough for the team simply to have accomplished an "upset."

But the impression was set, and there it has stayed for nearly a century. Crimson writer E. Benjamin Samuels asserts that "the 1921 Harvard-Centre game is remembered as one of the all time major upsets in the history of college football, and the Associated Press called it the greatest upset in all of sports in the first half of the 20th century."[37] Samuels elaborated: "Even if the Centre squad did not consider it an upset, most everyone in the outside world did, including and especially at Harvard. So yes, I would certainly call it an upset," a supposition with which many

Kentuckians, even in the first part of the twenty-first century, fiercely disagree.[38]

There is also the question of a rematch. Even today, some Kentuckians want this, and they believe sometime in 2021 would be great timing. Pure and simple, it's not going to happen. Says Crimson's Samuels: "I would consider it [a rematch] extremely unlikely. . . . The Harvard football program has expressed no interest, and the disparity in levels of play is far too great."[39] Some say that discussions were held between Harvard and Centre officials in 1971, the fiftieth anniversary of *the game*, and then again in the seventy-fifth year in 1996. Nothing came of these talks, "and the potential Harvard-Centre matchup was dropped." The two teams belong in greatly different leagues today. Harvard coach Tim Murphy remarked: "If we can occasionally get an Army, Navy, or . . . Duke, that would be great. But to play Centre wouldn't make a lot of sense. . . . Nothing to gain and everything to lose."[40] (Does this last statement imply a fear that Harvard *might* lose another game to Centre?)

Better, then, to leave things as they are. Even a Centre win in a rematch would somehow tarnish the luster of 1921. The story has frequently reached fairy tale proportions, with citizens from all walks of life giving their own interpretations and embellishing the events of *the game*, whether they were there or not. Former governor Happy Chandler said in 1977 that the 1921 game was "the finest hour in Kentucky football."[41] The boys on the team were fine representatives of the state, "decent on and off the field."[42] They were a great reflection of the student body at Centre—and, indeed, all the other state and private campuses in Kentucky as well. Applications to attend Centre soared after the victory over Harvard. Good high school graduates wanted to be associated with a college that not only had such high academic standards but demonstrated excellent athletic abilities.

George Morgan Chinn (or "Major Chinn," as he was now called because of his continued involvement with ROTC) received his college letter in 1921, allowing him proudly to display the big gold C on his shirts and sweaters. He still had injuries that kept him out of the games after the victory over Harvard. He "chafed at the bit" in 1922, literally begging Coach Moran to put him out onto the field. The coach always said he'd use Chinn only "if absolutely necessary"; as a result, though Chinn dressed out for each game, he saw very little field action.[43] Instead, he turned increasingly to coaching, carrying out Moran's directives and even suggesting certain plays and strategies to the coach himself. Increasingly, Moran listened carefully to Chinn's advice, with their roles of mentor-protégé becoming widely known throughout the football world.

A few weeks after the Harvard "upset," Centre won another football contest, but the publicity from it was not particularly welcome. Two "town" teams, Shelbyville and Pleasureville, were scheduled to play; it was widely forecast that Shelbyville would ransack Pleasureville. One of George Morgan Chinn's distant relatives coached the Pleasureville team. He telephoned George in Danville and asked him to bring some of Centre's players to participate in the game, which was to be played on the Shelby-Henry county line. George asked Moran about it, and Moran agreed, as long as the Centre players were not paid anything and there was no gambling. Moran thought it would be good practice for their scheduled game with the University of Arizona in a few weeks' time.

Seven Centre players, including George M. Chinn, were brought to the field with as much secrecy as possible. A crowd of some one thousand spectators awaited them. Amid cheers from both Shelbyville and Pleasureville, the game started. It did not take long for the Shelbyville fans to realize that they were not playing Pleasureville but Harvard-humbling Centre! One drunken Shelbyville fan, demanding to be let loose on the Centre players, had to be forcibly restrained. Apparently, much gambling went on during the game, with one source claiming "that a farm changed hands" and a Shelbyville bookmaker was wiped out. Not surprisingly, the win went to Pleasureville. Within a week the Southern Athletic Association (SAA) started an investigation of the Centre participation in this game, interviewing all seven players. Chinn alleged that "we were treated [by the SAA] like common criminals." All seven were exonerated when it was found that they made no money from the event; nor had they gambled. A few days later, Centre beat Arizona, becoming the national champions of 1921.[44]

Chinn was still officially on the football team, but Coach Moran would not let him into any of the action. He began to use Chinn more as an assistant coach than a player. Chinn was unhappy with this arrangement and perhaps that is what caused him to lash out at athletic programs at other colleges and universities in his column "Now and Then" for Centre's student newspaper. For example, he "humbly suggested" (of course, Chinn never "humbly" suggested anything) that the University of Kentucky change its mascots from Wildcats to "Raincrows," causing everyone at Centre to wonder what UK had ever done to Chinn. A rivalry in sports (baseball, basketball, and football) lay ahead between Centre-UK, and perhaps George was getting the student body ready. Chinn continued, "Every time they come over here [from Lexington to Danville] there is a deluge of rain and they take damper spirits back with them." Chinn argued that the raincrow makes a lot of noise and "really never does much."[45] Athletics and academics alike at UK made no response to Chinn's remarks—at least immediately. Chinn continued to enjoy his journalistic career.

When the 1924 season opened in late summer, Centre coach Charlie Moran stunned Centre as well as all other educational institutions in the Commonwealth by suddenly turning in his resignation to the athletic chief Myers. He had savored so many good seasons with the Praying Colonels and had enjoyed such good rapport with both the academic and athletic sides of the college that many people thought he was a permanent fixture, that there would always be a Coach Moran in Danville. All he would tell inquisitive reporters and even friends about the matter was that he was going up to coach another small college, Bucknell University in Pennsylvania.

A few days after Moran's resignation, there was another bombshell on campus. George Morgan Chinn announced that he, too, was leaving Centre, going to Bucknell as an assistant to Coach Moran.

2

Football and Caves

George Morgan Chinn considered several factors in his decision to leave the Centre College football lineup he so devoutly loved. He had traveled out of the state only once, when the Praying Colonels trounced Harvard 6–0. He treasured the memory of this trip to Boston and Cambridge and their environments, and obviously wanted to experience more of the world. The injury to his shoulder kept him always on the sidelines, which put him in the coach's company rather than the players'. Charlie Moran began to depend on Chinn, whose sharp eye for the field helped to gain Centre advantages. Chinn was not assured participation as a player in any future Centre football programs, and he worshipped "Uncle Charlie." Moran's departure meant a new beginning for Chinn himself.

Moreover, public opinion of Centre's football program had apparently slipped by 1923. Though it had nothing to do with the Praying Colonels, a rock-throwing incident helped to instill a paranoia that Centre's football program was not as welcome as it once had been. Details are somewhat unclear, but it seems that George Morgan and two of his friends, Minos Gordy and Rodes Ingerton, were on their way to Danville late one night when, about two miles from Gentry Lane, someone threw a large rock into their car.[1] George Morgan suffered a deep gash close to his cheek, which ultimately required several stitches. Furious, George and his comrades hurried to Danville, got a pistol and, somewhat foolishly, returned to the scene. George Morgan, still covered with blood, and Gordy and Ingerton beat up the rock throwers until they told them who was the instigator. All said Porter Shelly.[2] The case was ultimately settled out of court. The incident, however, soured George Morgan Chinn's opinion of both the citizens of Danville and the "gung-ho" supporters of Centre's football team.

Another incident that helped instill in Chinn a negative attitude toward the football program was when a "belle" from Memphis invited the entire team to her home for tea. (This was not unusual; the Centre football team was frequently invited to luncheons, dances, and theater parties.) Chinn wanted to accept the debutante's invitation, but team captain Edward Kubale ordered that a vote be taken on the matter. George Morgan very much wanted to go, arguing that "there's very little difference between a tea and a weenie roast, except the nourishment served." But the vote was 27–1 against going to tea with the Memphis debutante.[3] George Morgan Chinn did not like to be overruled, even by a definite majority; this was a trait he carried with him for the rest of his life.

After the victory over Harvard in 1921, even while he helped Moran coach the Praying Colonels on into the 1920s, George Morgan Chinn worked for a company named Lowe and Campbell.[4] His main job was to sell golf clothes and equipment throughout the state. Although not a golfer himself, he learned the rudiments of the game so he could talk intelligently with golfers, pro and amateur alike. He liked his life as a traveling salesman because it gave him the chance to travel, at least within Kentucky, and an opportunity to banter with people at country clubs and college and university campuses. He stayed in this position for about two years, until Coach Moran "rescued" him by asking him to go to Bucknell with him.

Another incident, the one that all but clinched the deal in favor of Bucknell, occurred on December 6, 1923, in Richmond, Virginia. The executive committee of the Southern Association of Colleges and Preparatory Schools (SACP) held a meeting in that city, and it was far from a pleasant experience. Inquiries and statements from some members led to raucous confrontations between the committee and Centre's president, Dr. R. Ames Montgomery. In short, Centre was accused of paying players to play football. Also, it was alleged, Uncle Charlie Moran was paid more for his services than the college's president himself. Dr. C. E. Allen, the faculty's chairman of the athletic committee, was outraged at these accusations. He read from a list to prove his points: when Coach Moran came to Centre in 1917, he coached football the entire year without pay; in 1918, he received the "huge" sum of $200; and in 1919 his salary came in at a whopping $500. Allen indignantly compared this to the salaries paid to head coaches at other institutions of higher learning, some of which had been among those condemning Centre for inflated costs: Vanderbilt, $12,000; Tulane, $14,000; Washington & Lee, $12,100; and South Carolina, $24,000.[5] After he cited those statistics, the shouting stopped and everyone went home, unhappy with the results of the meeting.

Back at Danville, President Montgomery was strenuously defended by George Morgan Chinn, who wrote a column for the local paper, the *Advocate-Messenger.* "Well, Prexy," he "wrote" to President Montgomery, "your carrying the ball in Richmond took them [the members of the executive committee of the SACP] on all together, tearing through them like the Royal Palm, Ltd., goes through Wilmore, and have won from all of us an unlimited admiration."[6] He referred to the leaders of SACP (which earlier had been the Southern Intercollegiate Athletic Association—SIAA) as "Saps in an Argument." The persons who gave the "incriminating" evidence against the Centre football team (Chinn called them "Birds") "couldn't have detected a barrel of fried onions" or "tracked a rhinoceros in a heavy snow."[7]

But the damage had been done; the tarnish stuck. Throughout the sporting world, fans began to wonder how Centre, with fewer than a thousand students, could field such a strong football team. There was no evidence that Centre paid athletes to attend the school, or that Moran was overpaid compared to other schools in the league.[8] (It is true that Moran was, in addition to his football duties, an umpire for the National Baseball League; even then his combined salaries did not match those of other college and university coaches.) Still, the disagreeable incident contributed to Moran's and Chinn's decision to leave Centre.

Neither man, however, cared much for Bucknell. Half (but only half) in jest, Chinn called Pennsylvanians "Yankees."[9] People there teased him about his Kentucky (southern) accent, mostly good-naturedly, but enough to pique his aggravation. The football team at Bucknell, though good, did not have the reputation of "giant killer" that Centre enjoyed. After a year or so at Bucknell, both Moran and Chinn left, the latter coming back to Kentucky to coach for a while at Wesleyan College in Winchester (the college moved to Owensboro in the early 1950s). Moran took a coaching position at Stanford University in California. There things might have stayed indefinitely—Moran at Stanford and Chinn at Wesleyan—if another tiny little college, this one in Salisbury, North Carolina, had not offered Moran the position of head coach of its football program. Again, Moran asked Chinn to go with him, and again, Chinn said yes.

Once again George Morgan Chinn was under the tutelage of Uncle Charlie, a most pleasing and auspicious situation for the young Kentucky coach. He joined Moran in Salisbury, where they would each take up football activities at Catawba College. George Morgan told an interviewer years later that Moran "made more impression on me than anybody that ever lived. Never was anything like him. We got scored on one year and by God, I thought . . . the man was going to go berserk. Scored on them!"[10]

George's grandson, Buddy Howells, told a magazine writer that George Morgan's experiences in football coaching taught him how to get a team ready for "operation," making certain "his players knew what to do, how to do it, and had the right equipment and training to get the job done."[11] This kind of mentality was of immeasurable help to him when he became involved in naval and marine ordnance, not least because he gained experience in anticipating how to troubleshoot problems before they even occurred.

Though they had arrived at Catawba just at the outset of the Great Depression in late 1929, which meant fewer players and equipment due to financial straits, Moran and Chinn piled up an impressive first season. They worked up enthusiasm among the team, creating a "scrappy eleven." In the 1930–1931 season Moran was late arriving in Salisbury. He was a baseball umpire for the National League from 1918 to 1939; his umping duties and some personal difficulties caused him to be late for the Catawba Indian practice sessions. During his absence, George M. Chinn in essence became head coach. "He is to be commended," the *Catawba Yearbook* exclaimed, "for his work and cooperation." Continuing, the *Yearbook* reporter said that "the first rudiments, necessary for a foundation, were taught by Chinn during Coach's absence." Chinn and his associates were among the men "who have made the 'Indians' the pride of 'Catub.'"[12]

The *Yearbook* actually complimented one of Catawba's opponents! While playing against Atlantic Christian College (ACC), the Indians faced a penalty for taking too long a time-out. When a spike came out of a local player's shoe, and it was apparent that even more penalty time would accrue, the coach from ACC ran out onto the field and earnestly asked the referee to forgive these incidents that led to delay of the game "because they were unavoidable."[13] The coach's request was honored. His team went on to lose the game, but he was hailed as a gridiron hero by Catawba fans, to say nothing of the ACC followers in the audience.

Chinn's viewpoint was that he reigned over his teams not only when they were on the field but in their private lives as well. In the past the football teams had eaten in the cafeteria at specially prepared "training tables." This aloofness from the student body as a whole was seen as enhancement; football players needed each other's company and all the good food they could get. In Moran's absence, Chinn put a stop to this practice: football players henceforth would mix with the other students, both men and co-eds. And, believing that the players had been getting rather too much of that good food available, he put them "on their honor" not to eat beyond their capacities. He and the entire college found that the honor system worked. A reporter observed one particular player for a few weeks and was astonished to see him, "meal after meal," push away food and desserts to keep fit for the next game. Why did he do it? the reporter asked: "Simply because he loved his team and cherished his word of honor." Coach Chinn had led this football player and his teammates in this direction, and his favorable reputation soared as a result.[14] No one was more pleased with this development than Moran himself. No wonder, then, that Chinn became BMOC ("Big Man on Campus").

By the end of the 1930 season (undefeated) Moran was back on the job and Chinn was once again his chief assistant. Nevertheless, it was Chinn, not Moran, who represented the football team at Catawba's annual Thanksgiving fest in November. Chinn spoke at this Catawba homecoming about the undefeated season just behind them, and how proud he and his wife, Cotton, were to be residents of Salisbury and connected with Catawba College. For all practical purposes he was now the head coach at Catawba. When Moran was on campus, he frequently called on Chinn for both strategic and tactical advice. Word got around to the coaches of the "Little Seven" that the coach at Catawba was "outstanding." Quite often, the coach they meant was George Morgan Chinn.

At a coaches' conference in Durham, North Carolina, in late 1930, George Morgan Chinn was chosen as a commissioner to study and make judgments of each player, not just those at Catawba itself but at all the other teams in the Atlantic Conference. Now Chinn had control and power, both athletically, to ensure physical fitness, and academically, to make sure each player's grades were high enough (usually a string of Ds would get the job done) to stay on the athletic roster.[15] He was now known locally, regionally, and nationally as one of the most accomplished football coaches in the entire country. He would not have been unhappy to have spent the rest of his life in this favorable capacity.

But then, as in the past, something happened that caused Chinn to travel in directions he had not anticipated. He and Cotton took a brief weekend holiday to a tourist attraction known as Bat Cave. (The famous poet and writer Carl Sandburg had a house nearby and, of course, they wanted to see this structure as well.) They had heard about Bat Cave from local townspeople, and since George Morgan and Cotton were from a state dotted with caves, they were naturally interested. They found this particular hole in the ground quite forbidding. A small lake within kept the ground constantly wet, and big boulders had rolled down into the cave, suggesting there were other caves beyond this one open to the public and making hiking or even walking hazardous. The guano left by the bat inhabitants did not help matters, either.[16]

As George and Cotton left the cave area, they did some quick comparisons. Formations of limestone rock on some land he had inherited from his grandfather, on the border between Kentucky's Mercer and Jessamine counties, a stone's throw away from the Kentucky River, were in better shape for tourism, the

Chinns thought, than Bat Cave. At some places, these Kentucky limestone palisades rose to 200 or more feet. Those on Chinn's land were some 150 feet in height. They were in the Brooklyn area, so named because a woman from New York who had moved to Kentucky years before had concluded that the bridge spanning the Kentucky River looked like Brooklyn Bridge: She renamed the bridge accordingly, and the community henceforth was known as Brooklyn.[17]

The idea for a new project was born. The Chinns' departure from Catawba was apparently not welcomed by the college's administration. The good citizens of Salisbury wanted him to stay in North Carolina, but, although George Morgan and Moran had had favorable seasons, the Depression continued to make inroads into salaries and equipment. This situation figured prominently in George Morgan's decision to return to his native state. "Unk and I had swell jobs," he told his good friend Happy Chandler. "He was head coach and I was Athletic Director, Asst." Chinn also claimed to be Salisbury's "weather forecaster," and he sang "outfield tenor" in the choir.[18]

There were other reasons Chinn was ready to leave. Though he did not come right out and say it, he, in fact, accused the person in charge of ticket sales (on which much of Chinn's and Moran's salaries were based) of graft and corruption. "At the end of the season, the Graduate Manager did more tricks with the gate receipts than a monkey could with ten miles of grape vine." He dolefully reported to Happy: "All we got out of last year was experience in large doses."[19] It was a bitter pill, but one that created a habit of careful watching whenever Chinn had dealings with either private businesses or government. After they left Catawba, Moran continued umpiring baseball and Chinn went on to new facets of his own life, which included not only Chinn's Cave House but work for several government agencies as well.

Back in Kentucky, preparing a cave entrance to his part of the palisades, George Morgan Chinn had an unpleasant experience that made him wonder whether he and Cotton should have stayed in North Carolina. One day, returning to his home in Harrodsburg, he rounded a curve in Corbin, Kentucky, and found the road blocked by policemen. At first he thought it was an accident, but a deputy sheriff explained to him that it had been a gun fight. One of the participants was a gas station owner from Corbin named Harlan Sanders.[20] Matthew Stewart killed Robert Gibson and Sanders seriously wounded Stewart. Little did George Morgan realize that in the near future, he himself would be the victim of violence, signifying that not everyone with access to the palisades welcomed him and the cave store he intended to build.

Chinn hired the services of a man named Tunnel Smith (Smith does not seem to have had a "proper" name, just "Tunnel"), who knew a great deal about explosives. Chinn, too, was an expert in explosives, but overwhelmingly of the military type, not the kind needed for this project. Smith began to blast away at the façade of the palisades with TNT, and soon he had a tunnel twenty feet long that went straight from Highway 68 to the back of the cave. Veering off to the left was another cave, judged by Smith and Chinn to be at least a hundred feet. Chinn was pleased; early on, he envisioned what he could do with this place in matters of merchandise. Even before the blasting was completed, George Morgan hung up a shingle at the cave's entrance reading CHINN'S CAVE HOUSE and began to implement his planned layout of the cave. The front tunnel had three parts: on the left as one walked into the cave, there was a twelve- to fourteen-foot wood-burning grill where Cotton made thick country ham sandwiches for sale at 15¢ each and "foot-long" hot dogs for about the same price; a few different brands of soda pop were also available.[21] In the middle was a walkway: customers could go to the end of the front tunnel but were usually forbidden entrance to the one-hundred-foot side tunnel. On the right of the front tunnel was a bar. Only grownup people could "belly up" to this bar, because Chinn built it high so that a child or even a teenager could not reach the top of it. Chinn wanted his customers to imbibe, even though the country was right in the middle of Prohibition (in fact, more booze was sold and consumed during this time than before Prohibition became the law of the land).[22] He did not, however, want to run a business that might encourage young people toward ruinous alcoholism. The high bar discouraged teen attempts to drink beer or other "moonshine" products. To help sales, George Morgan Chinn frequently placed half-eaten "sandwiches" (which turned out to be made of rubber, bought from a local novelty store) on the bar to create an incentive for customers to order food to go with their drinks—which would in turn induce a renewed thirst.

There were no chairs inside the cave itself; if one wanted to sit while enjoying a country ham sandwich or hot dog and "pop" (a euphemism for beer), there were a few seating arrangements just outside, between the maybe ten feet separating the cave's entrance from busy U.S. 68. Just feet across from the highway was the Kentucky River, which then had no line of trees shading it, as is the case nowadays. With cars and trucks zipping by on the road between Lexington to the east and Harrodsburg to the west, the quest for "comfort" was quite dangerous, especially after Chinn installed four gasoline tanks for drivers to fill up their cars or trucks. The outdoor seats were precipitously close to the highway itself; with motor vehicles speeding in for gas, coming to screeching halts before the tanks (gas was about 24¢ a gallon) it was sheer foolhardiness to keep sitting there.

But sit people did: to bring in customers George Morgan Chinn counted on his reputation as a football player on Centre's team in 1921, the magical year Centre "destroyed" mighty Harvard, as the well-known and liked assistant coach at Centre thereafter, and as the all-time winning coach at Catawba in North Carolina. He was well known in this area of Kentucky and many a patron stopped by simply to see Chinn and perhaps chat with him, lingering on the first bite of Cotton's delicious country ham sandwich. People flocked to his place by the hundreds or even thousands, keeping George and Cotton very happy indeed. Even people driving by who did not stop always gawked, curious, at the Chinns' unusual establishment. Though he had created quite a traffic nuisance for the Kentucky State Police, Chinn quickly became known as the "Cave Man of Kentucky."[23] He installed a sign at the cave's entrance: "If you Like Me, Come on in. If not—there's the River—go jump in it."[24] Many patrons smiled at this advice, attributing these words to Chinn's sense of humor. He kept insisting, however, that there was more than one meaning to the phrase.

An online writer remembered that in wintertime, a kinsman always stopped at the Chinn Cave House when traveling to Lexington and then back to Harrodsburg. She'd scoop melting ice off the palisades at the entry to the cave to mix with her glasses of bourbon, pulling off a sizable icicle to serve as a swizzle stick. Great crowds gathered between Lexington and Harrodsburg to "test the waters, especially during winter time, for their libations." They found the circumstances quite to their liking. An online oldster recollected going to the palisades and Chinn's establishment on his mule. A donkey, he asserted, was not good in bad weather. If it rained, sleeted, snowed, or iced during winter, a donkey invariably found a hole, made either by a rodent or bad weather (some places on Highway 68 were "mud puddles" after a heavy downpour), and fell down, being therefore of no use in getting its rider back home. So, said the memoirist, "if you're drinking" or "plan on drinkin'," always "ride home on a mule."[25] Cars, trucks, horses, mules, donkeys, and pedestrians almost always intermingled in the places around Chinn's Cave House, causing traffic jams and bad tempers.

Many patrons wondered how Chinn could make a profit by relying so heavily on country ham sandwiches and hot dogs. Of course, he couldn't. But what most patrons could not see was the completely illegal line of slot machines in the dimly lit left tunnel of the cave. Surprisingly (but knowing Chinn's local reputation, perhaps it is not so surprising), the authorities discovered the slots, charged Chinn with operating an "illegal game of chance," and hauled him into court. In typical self-assured Chinn fashion, he defended himself, arguing that the slots did not constitute an illegal game of chance because "you don't have a chance" when you gamble at Chinn's. Why? Because every slot was rigged so that anyone who played it lost money. The jury was convinced; he was acquitted. At the end of each day, Chinn opened the slots and counted the money. Although the machines were only for pennies and nickels, he still turned a goodly profit.[26] George Morgan finally collected all the slots in his cave, toted them across the road, and threw them into the Kentucky River. He was afraid young children would find the slots, start playing them, and become addicted. He did not want such a thing on his conscience.[27]

Later, when George Morgan was a Marine officer, one of the directors of a service club told him about his troubles with slot machines. They simply did not have very much money in them when he opened up for business each morning. He had heard that George Morgan Chinn "knew something about slot machines" and wanted his advice. He happily offered his help: "Well, there's an easy solution to that," he told the inquiring club man. "Just change the personnel who are emptying the machines at night." It worked![28]

The other major source of revenue at Chinn's Cave House, of course, was illegal liquor. Even in 1933, after the repeal of the Eighteenth Amendment, some county governments within Kentucky voted to remain dry. For some time, there was a gray area in Mercer County; it wasn't clear whether it was legal or illegal to sell alcoholic beverages. Naturally, George Morgan Chinn interpreted this situation in the most liberal terms he could. Everyone knew that beer, if legal, should be at least 3.5

percent alcohol. To bring all his stock up to par, George Morgan used "needle beer." He'd take a hypodermic needle and fill it full of pure alcohol. Then he'd pour off some of the brew on hand, or even use soft drinks, and inject the new beer into the old bottle until he knew his product was within proper limits.[29]

Eventually, Mercer County voted unambiguously to be dry. Chinn heard about it, but since he had no telephone in the Cave House and, actually, no "Rural Free Delivery," he considered that no official word had ever reached him about this newly imposed local prohibition. Even if it had, he argued that tourists from Cincinnati, being German, would not patronize a place that did not have ample supplies of beer for sale.[30] He considered it, he said, his "civic duty" to see to it that the Germans from Cincinnati were "well cared for." It took a while for officers of the law to discover his violations of the "local option." It took nine years, Chinn said, for the law to get word to him that his continued sale of beer (and other spirituous liquors) to "preferred customers" was illegal. "It took the sheriff and two deputies and a patrol wagon," he exclaimed, "to bring it [the word] to me then."[31]

As well as attracting the unwelcome attention of lawmen in reference to slots and liquor sales, George Morgan Chinn also had at least one unfriendly neighbor: fifty-year-old Price Peniston from Wilmore, Kentucky, ran a "recreational" camp on Highway 68 east toward Lexington. He had been in business for several years; besides selling groceries of all sorts, Peniston operated a camping ground. Tourists from Cincinnati, Louisville, Lexington, Nashville, and many places in between came to Peniston's camp for a weekend, sometimes longer, of rest and recreation. When Chinn opened his cave, the peace and quiet of the place disappeared—especially since George Morgan started up a firing range for himself and his friends in the back of the cave.

Relations between the two businessmen worsened until,

finally, these two rowdy ruffians did what a large number of Americans, both then and now, do: they let their guns settle the argument. In 1930 George, Cotton, and their three-year-old daughter, Ann, motored back from North Carolina for a short vacation in Kentucky. George had heard all along down in the Tar Heel State about the "bad-mouthing" coming from the Peniston side of the river. George drove up to the Peniston place, got out of the car, and "walked in the direction of a filling station" run by Peniston. As Chinn approached, "Peniston drew a gun and began firing at him." Chinn ran back to his car to retrieve his own firearm, whereupon he began shooting at Peniston. A newspaper reporter said that ten or twelve shots must have been fired by the estranged Kentuckians. Peniston was hit in the right leg and Chinn in the right side.[32] All along Cotton, who was known far and wide as *the* crack shot of Mercer and Jessamine counties, tried desperately to get George's firearm away from him, but he would not yield. He did not want this confrontation to grow any further, and he believed that Cotton would have killed Peniston right then and there if she'd gained possession of George's pistol.[33]

Peniston was conveyed to a medical facility in Wilmore. Cotton got George into their car and headed straightway for the A.D. Price Memorial Hospital in Harrodsburg. Their young daughter "had never seen so much blood in her life." Chinn got "patched up" at the hospital, but he never did get the bullet removed; he carried it for the rest of his life.[34] Chinn wisecracked even from his hospital bed: "After he [Peniston] had shot me three times in the belly [it was actually only once, in the right side], I decided he was mad. You know, I catch on quick that way." Both parties vowed that charges would be laid against the other as soon as each had recovered sufficiently from his wounds. As time passed, however, the incident was overlooked, never to be brought up again, except in the local folklore that inevitably grew from it.

A part of that local folklore was the so-called Sardine War. One day after George had recuperated from his "belly" wound, he grabbed his .22 rifle and walked down the road from the cave to Peniston's place of business. When Chinn entered the store, Peniston cringed as Chinn aimed right at him. It turned out, though, that Chinn was not particularly interested in Peniston himself. Instead, he leveled his rifle at a long shelf, reaching from one side of the store to the other, full of canned sardines (sardines were very popular among Peniston's patrons), and methodically shot each can off the shelf.[35] The Sardine War seems to have relaxed the tensions between the two; they lived thereafter in toleration of each other.

Increasingly, George let Cotton run the Cave House while he concentrated on building a firing range in the cave that branched off from his restaurant/bar. He had always loved firearms, at least those that could be used in military action. He lined the cave with thick steel plates so that even high-powered artillery could be fired into its back. Latter-day tourists found bullets scattered over the ground that were made of "some sort" of "lightweight alloy," with one side flattened out from striking the first metal plate.[36]

One day during an especially loud and smoky session of weapon firing, an attractive, well-dressed woman walked into the cave. She was taken aback by all the smoke, noise, and what seemed to her a great deal of confusion. She began to leave, but Chinn appeared before she got away. She asked, somewhat timidly, "Excuse me, Sir, do you still serve ham sandwiches here?" Chinn, noting her expression of anxiety, decided to inject a bit of humor into the conversation: "Yes ma'am," he replied politely and graciously, "we do, but we have just finished shooting the hogs and it'll be a while."[37] Without another word, the lady then *did* make a hasty exit from the cave, got into her car, and screeched out, heading for Lexington and civilization.

Another time when Chinn was in Frankfort meeting with

influential friends, including Governor Chandler, someone reportedly told him that a "fellow can get a girl in Frankfort for a ham sandwich," meaning one, of course, that Cotton made at the Cave House. George went home and told her the story. She rather impatiently asked him, "George, what did you come home for?" He answered: "I came home to kill a hog." Cotton did not always enjoy his sense of humor. Or whatever it was.

The couple kept running the Cave House in both its legal and its illegal aspects. Chinn once gave away a twenty-five-gallon still that he had secretly hidden in the cave to one of his regulars. Still, he gave every impression of being an honest businessman when, in fact, he was flaunting the hypocrisy of Prohibition and later the myth of "dry" counties in Kentucky. He certainly did demonstrate that the bootlegger's chief enemy was a legally "wet" county, for such would create unwelcome competition. Long after the Cave House closed in the late 1930s, George Morgan Chinn relished telling people that in operating Chinn's Cave House, he came to be "thought of highly in low places."[38]

He was proud of his political connections in Frankfort and of his increasingly important role with military people from Washington. By this time, it had become obvious to all but a few ideological diehards that the world would ultimately have to deal with the rise of dictatorships in Europe and Asia: Germany and Italy in the West and Japan in the East. Chinn's experiments with weaponry in his Cave House on Highway 68 between Harrodsburg and Lexington became of paramount importance as military ordnance men began to contact him about his activities with a view to determining how those activities might support future U.S. war efforts. It was a heady experience for Chinn to aid the administration of FDR and, truth be known, Chinn loved every minute of it.

3

ल

Odds and Ends;
or, Here and There

Predictably, once back in the Big Settlement, George Morgan and Cotton Chinn moved straightway to Mundy's Landing on the Kentucky River.[1] Many boats, both new ones and those in disuse, were docked at this popular river place. George and Cotton bought a one-hundred-foot, steel-hulled, Irish-manufactured Tyrone ferry and, now with George M. Chinn as the skipper, began to convert it into a huge Kentucky River houseboat.[2] They named it the *Iron Duke* and installed a small propeller on its stern that could move them along at a snail's pace to environs around Mundy's Landing.[3] Within weeks of his return from North Carolina, Chinn was deeply involved in several activities: getting Chinn's Cave House ready for business, commuting to Frankfort, where he performed several services for lieutenant governors, governors, and other high-ranking officials, and now restoring a ferry.

There was, however, one thing that, in George's opinion, couldn't wait. He simply could not abide being so close to the river without actually getting into it. Apparently, he seined all the way "up to his Adam's Apple" from Johnson's Bridge on the Kentucky to Clear Creek, which was a "mile past the Montgomery Place." All he wanted to do was bring in a respectably sized fish; instead, he woke up the next morning "with a taste in my mouth like a bilious pelican and a fever that ran the juice up in the doctor's thermometer like the motor meter on a frozen Ford." Continuing his hyperbole, Chinn avowed that his "temples throbbed like a mashed thumb, and my tongue bore a marked resemblance to those well known non-skid batter cakes that are commonly called waffles."[4]

He blamed his problems on just about everything and everyone except himself. He had gotten "soft," he said, and "it goes to show what . . . [effect] wearing underwear will have on a simple and untutored child of nature." He claimed that five years ago he could have "stripped off naked" and "climbed a thorn tree with a cub bear under each arm and never gotten a scratch." It made him feel as though he were a "sissy" to traipse all through Clear Creek (if it ever occurred to him to use an old-fashioned pole and line and fish from the bank, he never mentioned it) and then "turn up with one foot in the grave and the other on a banana peeling."[5] And, making matters at their absolute worst for George Chinn, he did not catch even a minnow!

He dolefully reported to Lieutenant Governor Chandler that he had "joined the ranks of the unemployed." Pointedly, he mentioned his father's recent visit to Frankfort and his conversation with Chandler—George Morgan Sr. had asked Chandler to find a job for George Morgan Jr. Chandler had promised to do the best he could. George Jr. sought to capitalize on this opening. He told Chandler, "Please don't put me in a place unless you are confident of my ability to serve," but he made it clear he would consider *any* job, except bookkeeping: he'd be about "as useful" in bookkeeping "as a glass eye at the movies." He could handle anything in the adjutant general's office, seeing as how he had six years already of military experience. (He was counting his years at Millersburg Military Institute as well as active involvement with ROTC while a student at Centre.) He had, he claimed, clocked up 1,784 hours of drilling from cadet private on to cadet major and then to the position of commandant. He'd love any position, he said, that required traveling, reminding

Chandler of the times when Chinn had roamed the Common-wealth on behalf of governmental programs and institutions. He closed this lengthy letter to Lieutenant Governor Chandler by reminding him that he "depended" on Happy being "right on the cuff."[6] A few days later, Chinn received a short and friendly letter from the lieutenant governor addressed to "Hon. George Chinn": "My Dear George: I will be on the cuff for you. Come to see me when you can."[7]

The friendship between these two young up-and-comers in Kentucky reflected the considerations and courtesies of the Chandler-Chinn families going back for generations. Jack Chinn seems to have been the last one of the family to dip into politics; when Chandler became involved, he could always count on the prominent Chinn family to support him with an enthusi-asm that extended even to modest campaign contributions. Both families were Democratic in their political outlooks and, with few exceptions, they championed the rights and privileges of all Americans to the equal enjoyment and responsibility of living in the Commonwealth of Kentucky and the United States.

Apparently, the first "cuff" Chandler seized for Chinn was to get him a job as a tour guide of the capitol grounds, a posi-tion that was at best part-time. Extraordinarily, there were three legislative sessions in 1936 in Frankfort; Chinn acted as sergeant at arms at two of them. After being nominated on January 2, 1936, by Senator Clarence E. Nickell of Kentucky's twenty-second senatorial district (Mercer County, among others), George Morgan Chinn was appointed by the senators as the sergeant at arms for the special sessions.[8] Senators spent their mornings in committees and did not meet as a group until noon each day. Consequently, Chinn had a considerable amount of free time on his hands. His chief responsibilities as sergeant at arms were to introduce important visitors to the senators while they were in session, keep senators mindful of their schedules if they should appear to be forgetting or overlooking them, control any tourist who might become "unruly" (that is, by interrupting a senator's speech or otherwise demonstrating unsatisfactory behavior, a situation that occurred only very seldom).

Chandler and subsequent governors and lieutenant governors gave George additional duties, perhaps in an effort to fill in more of his time when he visited Frankfort. His job as a tour guide of the grounds of the executive mansion was widened considerably to include much of the inside of the house. (There was even an unattributed rumor that some governors of Kentucky allowed George Morgan to live in a small apartment on the top floor of the mansion itself. This claim has never been proved or disproved.) As sergeant at arms and official tour guide, George came into contact with literally thousands of people, both influential and otherwise. He knew every senator and representative by name. He was acquainted with school superintendents, principals, and teachers throughout the state. And, quite important, he was familiar with just about every big corporate lobbyist who pitched his spiel (whether legitimate or otherwise) before Kentucky lawmakers. He was, rather unintentionally, getting ready for his later position as director of the Kentucky Historical Society. He knew everybody, and everybody knew him (a situation that he never tired of describing to his PhD "friends" around the Commonwealth). His circle of friends and acquaintances, soon to be augmented by military personnel from all over the United States and foreign countries, emboldened him in later years to call in favors. He had so many contacts that some of his associates actually hinted that he should run for governor; after all, he probably knew more about the internal workings of Frankfort than anyone else in the Commonwealth.[9]

Chandler's opinion of Chinn was always high. He liked George Morgan Chinn's "rambunctious" behavior. Chinn even turned his outrageous overweight to his advantage, meeting his critics with an "in-your-face" attitude. He began to tell a story that actually endeared his ideas of philosophical pragmatism

to thousands of people. He quit riding horses, he said, looking down at his sizable—and rapidly expanding—midriff, "for
humane reasons."[10] And certainly his friend A. B. Chandler—
lieutenant governor, governor, U.S. senator, baseball commissioner—continued to show his admiration for a man he once
thought he had killed. When Transylvania played Centre one
year, Chandler and another player, Rosy Dutt, tackled Chinn
on a kickoff. Chinn lay still on the ground for several minutes.
(Later, it was learned that he had a broken jaw.) As Chandler and
Dutt stood over Chinn, Chandler remarked: "We're going to
have a good day, Rosy. We've killed the toughest S.O.B. they've
got."[11] (This exclamation came before Chandler knew about the
longtime and solid friendship between the Chandler and Chinn
families in Kentucky politics, society, and sports.)

In one of his memoirs, Chandler heaped admiring praise on
George Morgan Chinn for supporting the success of Chandler's
lieutenant governorship. "From the time I was elected Lieutenant Governor, I usually had an assistant around you might
term a bodyguard." The first one, he said, "was George Morgan
Chinn, whom I originally saw as a tough little river rat at Mundy's Landing." George Morgan Chinn would have been infuriated at anyone else calling him a river rat, but when his friend
Happy Chandler said it, however, he took it as a compliment.
"He was strong as an ox," Chandler avowed, weighing "about
three hundred thirty pounds." When Chandler's "enemies" (that
is, those who did not agree with Happy on various points and
issues) criticized him, if Chinn were either present or nearby,
"these fellows" didn't take any chances with him. "They weren't
afraid of me, but they were afraid of him."[12]

George Morgan Chinn would see to it that Governor Chandler was safely tucked in at night; he'd come to Chandler's bed-
room and get him up in the morning, make sure he had breakfast
(whether he wanted it or not), and then walk with him to the
capitol and stay with him during the day. "He did as he pleased,"

Chandler asserted. "Ran Easy. In fact, nobody made him do or not do anything. Had him nearly all the time I was governor first time."[13] He was usually on the grounds of the executive mansion and capitol, at least through legislative sessions. When things were relatively quiet, he'd slip away to Harrodsburg, to Cotton, and to Chinn's Cave House, for which he had momentous future plans.

For many years, George Morgan Chinn tried to get friends and acquaintances to call him "GM" or maybe "GMC," and sometimes he even wore caps and shirts with the monogram GMC written on them in broad letters. His persuasions came to naught although he was approached by several individuals who wanted Chinn's help in obtaining a loan from the General Motors Corporation so they could buy a Chevrolet or some other GM product. After a while he discontinued his efforts; after all, the family had changed its name once, from des Chynn to Chinn. Better, he felt, to leave it at that.[14]

As he continued his duties in 1936 as the Senate sergeant at arms, tour guide, and all-around "handyman" both inside and outside the governor's mansion, Cotton was left behind to tend to Chinn's Cave House. Business had slowed considerably by the mid- to late thirties; there was enough revenue coming in, however, from George's work in Frankfort and Cotton's secretarial services at various places of business, including, at one time, Centre College. Also, the Chinns continued to sell calcite from off the land George had inherited from his grandfather John Pendleton Chinn. The Chinn family was in no danger of poverty as Chinn's Cave House declined in popularity, although Chinn did continue and even augment its use as a noisy firing range for several weapons he owned. It closed to the public in 1938 and lay dormant for the next twenty-two years, when Chinn reopened it—but for reasons other than selling food and drink.

In fact, George and Cotton were still so well off that they

began to build a house of their own on top of the palisades, from which they could see the entrance to Chinn's Cave House. All the noise emanating from building the house, plus periodic sessions at the firing range in the cave itself, caused some neighbors to register complaints. George was accused of bringing harsh noises to the otherwise peaceful vicinities of the palisades. The neighbors had a point, but that did not stop George, especially since he now spent most of his time in "faraway" Frankfort, and could not himself hear the cacophony.[15] The neighbors be hanged—he knew what he wanted, felt he should have it, and was bound and determined to get it.

The house was constructed of limestone gathered from his own property atop the palisades; consequently, no transportation devices of any sort were employed getting the stones to their proper and desired locations. Chinn employed workers to assist with the operation. He and workman Charles Nichols went out to the "quarry" almost daily to pry loose slabs, which could sometimes easily weigh six tons or more. Then they were lifted up by a power winch or, as the workers called it, a homemade overgrown tractor. These were taken to a nearby shed, where there was a large, circular saw with which David Lyons noisily trimmed the slabs to workable sizes. After this process, the stones were sent on to Bill Reynolds, who operated a mechanical buffer. He smoothed and polished them over and over, sometimes for a whole day, until they met the standards of appearance and usability demanded by the workers and George M. Chinn. It was a long, tedious, and difficult undertaking. But to have transported these slabs without trucks or wagons was truly astounding. Chinn accessed the tiles where he found them and displaced only those to be built into his residence. He left the others unscathed, at least until he began to sell some of his stock to nearby friends and neighbors.[16]

Asked whether some brick would be installed—as was the case with so many other limestone houses in Mercer County—

Chinn showed his architectural "purity." One could, he admitted, put brick into an otherwise limestone house, "but you ought to be arrested if you do." Many compared the building of the Chinn residence with the methods and materials used by ancient Egyptians when they built the pyramids. Many times the Egyptians used rollers made of smoothed-out trees to get stones to their desired locations. Was this somewhat equivalent to the rail system installed by the workers at the Chinn house to get the great slabs from one workplace to another? Chinn and several others argued yes. He said that the biggest difference between methods used in ancient Egypt and in the erection of his own house "was the jackhammer."[17] The house was completed in 1939 and became one of the most notable houses in a county filled with notable houses.

Each room of the Chinn house was filled with antique furnishings, relics, and mementos that George had inherited from his uncle Kit and from his grandfather Jack Pendleton Chinn. "We built the house around the furniture. We had acquired so much and just didn't want to get rid of anything." They even determined the size that each room of the house would be by measuring "the amount of furniture that would have to go in each."[18] The door leading to the living room was nine feet, five inches high, eight feet wide, and four inches thick. It weighed a bit over seven hundred pounds, requiring ball bearing hinges so that it could be opened and closed with relative ease. The house's interior had ceilings twenty feet high, red cedar walls, and a huge stone fireplace. An enormous chandelier was connected to the ceiling by heavy chains. The light fixtures on the chandelier were "from the first Pullman train car shown in the United States at the New York World's Fair."[19]

Almost immediately word got around in Harrodsburg and surrounding areas that Cotton and George's new house had machine guns mounted on its roof, manned by twenty-four-hour surveillance teams there to protect the residence. It was asserted that the Chinn Cave House down below the hill was also protected in this manner.[20] No definite proof of the truth of these assertions has ever surfaced. Perhaps the general thought that machine guns were bound to be everywhere on the Chinn property had some effect on discouraging unlawful visitation.

Nonetheless, trespassing at the Chinn house was not uncommon, chiefly when he and Cotton were away, which was often. They may have finished and furnished the house in 1939, but it was another twenty years (1959) before they lived in it permanently.[21]

During the twenty-year interval between the house's construction and George and Cotton's permanent residence in it, George M. Chinn still had the wanderlust. George was often abroad, serving in World War II and Korea. If not in foreign places, he was in Frankfort, serving whatever governor was in place and watching over the grounds of the capitol.

Meanwhile, a New York movie company, Rochemont Productions, contacted George and Cotton with regard to making a documentary about the evolution of bourbon whiskey, to be called *The Happy Age*. All the players in this cinematic endeavor, both amateur and professional, were Kentuckians. Colonel George Morgan Chinn was asked to play the part of the bartender. Though he no longer imbibed, he readily agreed to this role and, despite his daily responsibilities for the Kentucky Historical Society (he had been named director in 1959) in Frankfort, he allowed the front room of his house, with its huge fireplace, to be the setting. A twenty-one-foot-long bar was installed, laden with cheeses, breads, and apples. There were at least eight "testers" on hand to try each bourbon as it was presented to them. Colonel Chinn was one of them! None of the testers actually drank any of the whiskey brought before him, however. A taster might "wallow" the liquid around in his mouth and then expel it through the nose or spit it out, or he might rub a goodly quantity of it on his hands and smell it. Referencing this technique, which went back to pioneer days, Chinn remarked, "If the whiskey smelled exactly like a ripe apple, it was considered top quality."[22]

In the late 1930s Chinn, like most pensive Americans at the time, worried about events around the world. He was certain that sooner or later the United States would be drawn into the European and Asian conflicts, which seemed to intensify on a daily basis. He took a U.S. government weapons consulting job at Frigidaire in Dayton, Ohio. Here he worked in the weapons section on the .50 caliber aircraft machine gun. The war that eventually broke out in 1939 was one, he very well knew, that would have worldwide repercussions, and he hoped that his efforts at Dayton would help to fend off Nazi advances against England, France, and other Allied countries. It was the fear of a worldwide conflagration that took George Morgan Chinn to see his old benefactor, Happy Chandler, who was now serving as one of the two U.S. senators from Kentucky; he was on the Senate Foreign Relations Committee.

That the Chinns and Chandlers were still extremely close not only in Kentucky politics but in regional and national societal circles as well was evident in a letter George wrote to Happy in late 1937, when Chandler was still in state government. First, he told the governor, "I have been as busy as a germ at an epidemic." From what he had read in newspapers, this last legislative session for 1937 would be as exciting as "an old maid playing squat in a bed of asparagus." It was apparent to Chinn at least that Chandler would not actually need him during this upcoming session; everything, he predicted, would be quiet. Therefore, he had decided not to run for full-time sergeant at arms for the state Senate. "I may go to Florida," he told the governor of Kentucky, "as I need to lose some weight." He was perilously close to the 350 mark, and it affected him both personally and publicly. "It's gotten so my power over women consists of using the penny scales motto on them (Honest Weight—No Springs)."[23]

Then George Morgan segued into the sort of writing that he obviously most enjoyed, since he did so much of it: that of hyperbole. He was not, he said, the "gushing" or "flag-waving type," yet he was "still grateful and ever will be" to Chandler "for the many favors past and present." In closing this letter, he wanted to assure Happy and Mildred, his wife, that "if at any time either of you want a neighbor's barn burnt, or his well poisoned, a three-cent stamp on my address will get the job done."[24]

Of course Chandler knew that Chinn was joking. In fact he had no barns to burn or wells to poison, but it was nice of George, after all, to offer such services. Chandler, however, should have fully realized that this letter meant a request would soon be on his desk. It was George's habit to speak to possible benefactors in charming and complimentary ways before hitting them up for the job he wanted. He was never more accomplished at this tactic than with Governor Chandler, who moved up to Washington, DC, shortly after he received Chinn's letter, as the new junior senator from Kentucky. Chandler was quickly put on the Foreign Relations Committee, and it was there that George M. Chinn met him—or, as some of his critics argued, rigorously confronted him.

Citing his Marine experiences at Millersburg Military Institute and at Centre College, Chinn literally stood before the senator in Washington and "demanded" that Chandler use his influence to get him back into active duty. At first, Chandler could not believe what he was hearing. "Go home, George," he practically shouted at his visitor. Close personal and familial friendship was one thing, but manipulating the military system for a private favor was another. "You're over forty years old" (actually, he was thirty-eight) "and you weigh 330 pounds! You'll get in the way and you'll get killed, and you'll get other people killed, too. Go home!"[25] Chandler asked Chinn what he thought he would do in the armed forces, anyway, either in the Marines or the navy. He found it difficult not to laugh at his friend when

George answered, "Tail gunner."[26] Although everybody knew that George was a weapon expert, Chinn was just as aware as Chandler that there was no military plane in any branch of the armed forces where Chinn would be accepted as a tail gunner. George's outrageous assertion may have been the key to Chandler's agreement: if George wanted in the military so badly that he was willing to seriously injure himself to do so, who was he, Chandler, to deter or stop him?[27]

George seemed to understand his "adversary" was wavering. Chinn just kept standing in front of Senator Chandler, sometimes even at attention, not giving any ground. "No," he kept saying to the senator; "I'm gonna stay here with you and you're gonna look out for me until you get me back in the Marine Corps." Then he threatened to move in with Senator Chandler and Mildred. "Oh Lordy Lordy," exclaimed Chandler, knowing from experience that a Chinn in his household would create considerable culinary and economic chaos. "I can't afford to keep you," a resigned Chandler finally said. "I'd have to board you out because I can't feed you."[28]

George did have an ace in the hole, so to speak, that he thought he might be able to use if Chandler turned out to be made of stone on this issue. George had tried to enlist in early 1939, when the "recruiting train" had come through Harrodsburg, "taking applications" for military service and collecting résumés.[29] After several weeks, giving up on a positive development deriving from the recruitment train, Chinn went to the recruitment office in Louisville to plead his case with the recruiters in that city. The head of the recruitment office, Colonel Francis Kieren, knew of Chinn's knowledge of guns and his talents. In effect, he hinted that if Chinn could get Chandler's approval, he (Kieren) would support the necessary waivers to get Chinn into either the army or the Marine Corps.[30]

Shrugging his shoulders, Chandler picked up the telephone and put in a call to Quantico down in Virginia. He finally connected with Commandant Best and told him that Chinn was in his office, and would he kindly "come and get him."[31] Best, who knew of the obstreperous qualities that Chinn was capable of displaying, replied, "All right." So Best, with a few Marines beside him, drove to Washington and "got him." Chinn surpassed the age limit for the Marines by about ten years and the weight limit by some 118 pounds.[32] Chinn's case was undoubtedly the most significant waiver given by the Marine Corps during the entirety of World War II, or maybe even U.S. military history.

During the next several weeks, George endured what can only be called basic training, the kind enlisted men go through. At Quantico he shed some fifty pounds but was still dreadfully overweight. "A lot of my friends," he remarked wryly, "were coming out [of the Marine Corps] when I was coming in. To say boot camp was difficult for a 330 pound, 38 year old, would be a gross understatement." Of course, all his comrades were younger than he, and they considered it "an insult that the old fellow was still living at the end of each day."[33] After a while at Quantico he was transferred to Fort Knox, not too far from home in Kentucky, where he continued to lose weight. His aptitude tests placed him high in experimental weaponry equipment, both ground and aviation. He was again transferred, this time to Patuxent, in Maryland, where he worked at "Cuckoo Academy" and the "Screwball Institute" to troubleshoot U.S. weapons difficulties everywhere in the United States and throughout the entire world, if need be. Thus began a career (or, as George liked to call it, a resumption of a career) that would last for the next three wars, from World War II to Vietnam. (Some of his equipment, however, such as the M-19 automatic grenade launcher, made it further—all the way to Afghanistan in the early twenty-first century.)

If anything, the process of getting George Morgan Chinn back into the active military increased Chandler's respect and fondness for him. He heard later from Chinn that the training they gave him at Quantico "damn near killed him." After two weeks, however, Captain Chinn was in stride with Marine military discipline. "He was one of the toughest men I ever saw in my life," Chandler admiringly reported. In one of Chandler's conversations with J. Edgar Hoover, the FBI man reportedly told the senator that Chinn was one of the "best automatic weapons [experts] the Army had" (although, actually, Chinn was in the Marine Corps). Chinn would go on to serve with dignity and distinction throughout World War II. Chandler summed up his friend simply: "George was a remarkable fellow." Any descriptions beyond that affirmation would have been superfluous.

4

∾

Semper Fi

The more Captain Chinn struggled with weight issues (at one time in his career, he had to buckle two belts at his waist to keep his pants up), the more his good intentions fell apart.[1] At one base he lost seventy pounds, only to gain it back at the next post. He joined enlisted personnel in what was essentially basic training at Quantico in Virginia, some forty miles south of Washington, DC, climbing obstacle courses, taking hikes, crawling on his belly, and the like. He quipped one time that the recruits, all of whom were much younger than he, seemed to be somewhat disappointed that the "old man" was still living at the end of the day. He was under military orders to lose weight, but each time his weight fluctuated the size of his uniform changed—at the military's expense. Finally, Chinn decided to leave it alone, to stay at the weight that seemed most comfortable for him (he usually hovered around three hundred pounds). Medical doctors frequently stopped him out on base, thinking he was a plant, put there by their superiors to make sure they were on their toes in spotting unhealthful practices among the men under their command.[2] Chinn showed up at one doctor's office after being told to "be there." As the physician looked Chinn over, he remarked to one of his colleagues: "I knew they [recruitment offices] were scraping the bottom of the barrel, but I've got the barrel right here."[3] In the years that followed, Chinn liked to relate this story to audiences. He described himself as "the despair of the medical profession" because their predictions about him had been wrong: "The Navy's doctors are sore at me because I've upset their statistics chart. According to it, I was supposed to be dead 12 years ago last month."[4]

Unbelievably, Chinn let the statement by the sarcastic naval physician ride without any response. He really *did* want to get into the army or the U.S. Marine Corps. He knew that if he quarreled with an established military figure, particularly someone who outranked him, he'd come out of it negatively. So, unusually for him, he kept his mouth shut.

There was another matter on which he remained silent. According to press reports, he and the U.S. government made an agreement by which twenty-two hundred acres of Chinn land in Mercer County (which he had either inherited or would in the future) was planted in marijuana (or "industrial hemp") to be used as emergency rope-making supplies for naval and other military purposes. The permit for such cultivation was posted in his office, in big letters so that no one could possibly miss it. Even so, Chinn remarked, "Many . . . FBI men . . . nearly climbed over the top of my desk to see that permit." Years later, Chinn speculated how much a marijuana crop of twenty-two hundred acres in 1940 would be worth in 1983: "You'd be able to pay off the national debt," he quipped.[5] It was, of course, practically impossible to conceal fields of growing hemp in Mercer County, as in other places throughout the country. Many curiosity seekers, and sheriff's deputies as well, visited on a regular basis.[6] The easiest access to Chinn's marijuana fields was by boat on the Kentucky River; even so, one would have to climb the palisades to get to them. No one, not even Chinn as the landowner, was allowed on the property without government permission.[7]

One of the earliest users of marijuana was General George Rogers Clark. After he fell into a large fire and had to have a leg amputated in 1809, the "weed" eased his pain. The effect of marijuana was not lost on the early Kentuckians; however, they usually favored whiskey. The stalk of the plant was used for rope to help supply the navy; the leaves and buds were smoked by some citizens—at least when they could not acquire any bourbon. "I've seen people leave a field of hemp," said Chinn, "to go fire up the still." Most hemp was grown in the Bluegrass region; George's father, in an earlier age, had also planted sizable quantities of hemp, passing the techniques on to his son. Thus, it may be said that George Morgan Chinn, in raising hemp, merely carried on a family tradition. There was one great negative—at least for some individuals—in the growth of hemp in the nineteenth century. When a convicted criminal was sentenced to death, the authorities would obtain hemp rope from a nearby "ropewalk," the name for a hemp factory, and fashion enough rope for a "marijuana" hanging—or "hangin'," as the public usually called it.[8]

After Quantico Chinn went to Fort Knox for a while, and then transferred to Cherry Point, in North Carolina, an assignment that must have been nostalgic, since he and Cotton (who stayed at home in and around Harrodsburg during Chinn's military tenure) had spent some happy times together at Catawba College when he coached the Indians. At Cherry Point, Chinn spent much of his time as a wing gunnery officer.[9] Even before the war, Chinn had helped to turn the Frigidaire Company in Dayton, Ohio, into a maker of Browning .50 caliber machine guns.[10] At Cherry Point he continued experimentations with machine guns; ultimately, through trial and error (during which he endured serious burns on his hands), he came up with a "booster" for the .50 caliber, "also called a 'blow-back adapter' which doubled the rate of fire." Several colleagues marveled at this accomplishment, especially because Chinn had not received any kind of engineering degree at college. "There is no such thing as formal education in ordnance. With formal training you can study to become an 'engineer' in almost anything." In ordnance, however, Chinn asserted, "It's a case of aptitude more than any other factor."[11]

Chinn was gratified that with the eleven improvements on the Browning .50 caliber machine gun (he did not build a new machine gun; he simply added things and took away things that he thought would make it work better), he more than fulfilled the mission of the U.S. Bureau of Ordnance and the U.S. Bureau of Aeronautics, both of which operated in World War II and the Korean conflict: "To plan and develop by years of experience the needs of Naval Aviation and base its requirements under conditions of war, and never upon those of peace. The principal function of this organization is to harness authority and responsibility so that they can never be separated."[12]

Many officers with higher rank than Chinn were impressed with his work. They planned to send him overseas—to the Pacific theater—where ordnance problems hampered U.S. war efforts against the Japanese. Other officers felt that Chinn's greatest contributions would be at home; he should roam around the country to one military establishment after another and review recently manufactured and reconditioned weaponry. Equipment could thus be fully active before it left the United States. This procedure, argued many—a thought with which Chinn agreed— would save time, and lots of it; it would be much more efficient than chasing down malfunctioning guns on some remote island in the Pacific. His superior officers thought otherwise; though he disagreed with an overseas assignment, he was still a Marine, and he had to obey orders. He was commanded to go to the southwest Pacific, where he would be in the proximity of literally hundreds of islands where the war was being fought. His job was to respond to one application after another for someone to get a gun to function and keep it that way.

Then a Marine Corps chaplain with whom Chinn had made friends at Cherry Point intervened. The Reverend Charles H. Blumquiest had expressed interest in Chinn's work on the Browning .50 caliber machine gun. At the detachment's San Diego staging area, Blumquiest appealed to Marine Corps veteran Major H. Ross Jordan, "There's a big Kentuckian in my outfit about to sail." Explaining Chinn's work on the machine gun and other weaponry, Blumquiest highly recommended that he be allowed to bring Chinn into Jordan's office and talk over some things. Jordan started the conversation with Chinn: "[I] understand that [you] have something . . . to talk about." This query must have nettled Chinn, for he shot right back at Major Jordan: "I don't talk; I shoot!" He explained his work at Quantico and Cherry Point. When Chinn finished, the major picked up his telephone and put in a call to Brigadier General Ross E. Rowell, asking that Chinn be separated from his present unit and that he and Jordan be sent to Washington, DC, where Chinn's information could be presented to high-ranking ordnance authorities.[13]

Chinn was made available for consulting and offering advice on gun matters, and these requests came from all over the country. He was most immediately assigned to the Naval Air Test Center at Patuxent River, Maryland, where he made suggestions for improving some of the weaponry then being built, sometimes even supervising the work.[14] He also worked for a time at the Navy Proving Grounds at Dahlgren, Virginia; the Naval Aviation Ordnance Testing Station at Chincoteague, Virginia; and at the vast China Lake expanse at Inyokern, California.[15] He traveled mostly on military aircraft; he flew so often that the government issued him a VIP boarding pass that enabled him to bump other passengers if seats were filled. (And woe betide any passenger unlucky enough to be seated next to the hefty Colonel Chinn.)

On one occasion, however, this procedure did not work out the way Chinn wanted it to. The plane was full and Chinn and another person vied for the seat of a hapless passenger about to be bumped. High-ranking officers looked over the boarding passes of each and gave the seat to the other person. It turned out that he worked on the Manhattan Project and was personally involved with the atomic activities of J. Robert Oppen-heimer, whose mission—at least at the moment—was of more importance than Chinn's. Later, when word of this incident got around in the military community, the big quip was almost always: "Well, what do you know? One 'Fat Boy' was 'bumped' off a plane seat by another 'Fat Boy.'"[16]

Another story, almost certainly apocryphal but still very much in character with Chinn's penchant for practical joking, occurred when he was in Long View, California, on a weapons assignment. Part of his duties was to march troops up and down main thoroughfares in host cities. His outfit that day was composed mostly of New Yorkers and New Englanders. Chinn's acute eyesight caught a Confederate monument down the boulevard. Ordering, "Eyes right!" to the troops, he told the bandmaster to strike up "Dixie."[17] It was all in fun from Chinn's point of view, but some of the northerners did not quite see the humor in it.

Chinn did make it overseas. Just as the war was ending in 1945 (and after the atomic bombs had been used), he traveled to Japan: to Tokyo and a few other cities. He and his crew inspected the ordnance of the occupying forces to assure that any uprisings could (and would) be met with efficiency. In many instances his very presence on an island in the Pacific theatre was confidential; if the U.S. government knew that certain weapons needed maintenance or repair, the fear was manifest that the enemy would know about it as well, and possibly attack.

By early 1946 he was back in the United States, at Patuxent Naval Air Test Center, where he had spent most of the war itself. He luxuriated in a letter of commendation from Rear Admiral William Brown of the navy's Office of Research and Inventions. Chinn, said the admiral, had been Patuxent's project engineer on a number of applications of military weapons development and maintenance. His work with various aspects of the Browning .50 caliber mount, along with the twenty-millimeter cannon mount, "has been of substantial value to the Navy." The admiral said that the success of the production models, the ones that worked in military settings, "is in great part due to the design of the prototype [which Major Chinn and his team put together] and it is considered a highly satisfactory achievement."[18] Secretary of the navy James Forrestal lauded Chinn for "initiative, constant enthusiasm, unselfish devotion to duty and outstanding professional skill . . . in keeping with the highest traditions of the United States Naval Service."[19] FBI director J. Edgar Hoover joined the "Chinn Club," admiring the changes the colonel had made to improve the Browning .50 caliber machine gun (John Browning was this gun's inventor; Chinn added to its design and effectiveness in several ways). Hoover remarked that FBI laboratories would find Chinn's book valuable "in connection with the examinations of firearms and ammunition."[20] Chinn still weighed over three hundred pounds, but at this point in his career, no one, not even naval physicians paid any attention to his obesity. It was clear that Chinn was hitting the "right notes" in Washington.[21]

The U.S. Navy commissioned Chinn and a team of writers in 1946 to research and write a four-volume work called *The Machine Gun*. This ambitious undertaking kept Chinn and a crew of researchers busy for the next several years, with volumes 1 and 2 coming into print six years later, in 1952. These volumes amounted to a half million words and 417 illustrations. A reviewer, F. W. Foster Gleason of *Military Library*, lauded the two books, first for their "historical" approach to the machine gun, and second for emphasizing theretofore neglected ordnance as the "crux" of all warfare.[22]

Volume 2 argued that the U.S. Civil War, 1861–1865 (though Chinn referred to the conflict as "the War between the States"), was the impetus for many strong European powers to replace their conventional weapons. Russia, for example, was almost "totally dependent on foreign sources for its infantry armaments." The book reported that in 1871 Tsar Alexander III sent General Alexander Gorloff to Hartford, Connecticut, to study the Gatling gun; he bought four hundred such guns from the U.S. factory and distributed them throughout the Russian empire. The guns were stamped "Gorloff," which caused many people, including military personnel, to believe he invented or manufactured them. Thus, for many years, these guns were known simply as Gorloffs, without any explanations of their true origin.[23]

In October 1928, the First Five Year Plan was instituted in the Soviet Union. Chinn and his fellow authors wrote: "This plan, though widely heralded as intended to improve the lot of the masses through industrialism, was actually a thinly disguised scheme to establish a strong armament industry." This plan, the writers asserted, was accompanied by "warlike slogans" that emphasized "the hostility of the outside world." Of particular chagrin to that "outside world" was the fact that since Communism supposedly did not recognize property in reference to outside inventions and developments, "many of the world's inventors saw their ideas used by the Russians without compensation."[24]

Most civilians in the United States, or anywhere else in the world for that matter, not only did not read these two books; they did not know they even existed. Soviet officers knew about the books, but they were not able to read them because they were stamped as "Confidential" when first printed, and thus the Soviets were left with suspicions and suppositions about Chinn's "bellicose" remarks. Even before the first ten pages of volume 2, Chinn had succeeded in "insulting" several military (the "Gorloffs") and civilian (Five Year Plan) activities put in place by the Soviet government. It was not until July 1970 that the books were "downgraded" to "Unclassified." Of course, since many Americans and Soviets *knew* the books existed but not what they said, these uncertainties added to the heated rhetoric of the cold war in years to come. Many suspicions were generated by being left in the dark, so to speak, about the true nature of the opponent's weaponry systems.

The truth be known, Colonel George Morgan Chinn in 1946 loved being in the U.S. Marine Corps. He was a high-ranking officer, greatly respected for his abilities in ordnance, and had been granted what he considered the opportunity of a lifetime: charged with the duty of writing a history of U.S. Navy ordnance. He spent much of his time at Patuxent Air Test Center in Maryland and, just a short distance away, in Washington, DC, where most of the records he needed for his studies were located; and, if he wanted to get away for a while, he'd go to Harrodsburg, where Cotton took care of their interests at home. It took him the better part of a decade to write and publish volumes 3 and 4, and these—since they dealt with the ongoing arms race with the Soviet Union—were placed on High Confidential status by the U.S. government, which meant that only the highest of military personnel could read or even consult them. They, like volumes 1 and 2, were finally released to the public in 1970. Even so, the price remained high: today, for only one volume of the set, the price is $100—if one can even find a copy. Notable *Courier-Journal* columnist Byron Crawford called the first four volumes of *The Machine Gun* the "Genesis, Exodus, Leviticus, and Numbers" of automatic weapons.[25]

World War II ended in Europe in May 1945, and in the East in late summer; a scant six years later (1951) the Korean "police action" began. Colonel Chinn could have taken civilian status at the end of World War II, but when the time came in the 1950s, he exuberantly signed up for another hitch with the U.S. Marine Corps. He wanted to continue working on *The Machine Gun* and to develop new weaponry and refine the old.

Altogether, Colonel Chinn and his crew were responsible for several inventions, developments, and renovations of weapons during World War II, Korea, and Vietnam.[26] One of the earliest, besides the Browning .50 caliber machine gun, was the twenty-millimeter turret, used on tanks along with other rapid-fire weapons. The project itself started in February 1945, only a few months before the end of World War II. Naval lieutenant Frank E. Wood Jr. wrote to Chinn about contracting an independent company to do the preliminaries to relieve the work load at the Armament Test Shop at Patuxent. After initial planning, the lieutenant wanted an "A" priority for a full development of the turret. Since the turret was Chinn's idea to begin with, Wood wanted him "on the project as much as possible."[27] The improved turret was used primarily in Korea and Vietnam.

The United Nations "police action," more accurately known as the Korean War, raged from June 1950 to July 27, 1953, when it ended with an "armistice" on the 38th parallel. Colonel Chinn was assigned to Korea and spent some time in Seoul and in Tokyo. While Chinn was in Korea a MiG pilot deserted and surrendered himself and his plane to the Americans. Chinn was called in to inspect all the firing mechanisms on board. What he found was unremarkable by comparison with American aircraft, but it did enable him to learn many MiG patterns that had not previously been present in the U.S. system.[28] His greatest problem with the MiG, as with all other aircraft he boarded, U.S. and otherwise, was his weight: it was difficult for him to get in and out of planes. But he persevered.[29]

While he performed maintenance duties for soldiers in the Korean field, he also began work on what ultimately became his "pride and joy," the M ("Mark" or "Model") 19 automatic grenade launcher, which was used extensively in the Vietnam conflict and today (2014) in Afghanistan. He was on active duty in Korea when he and two of his military colleagues, Walter R. Cashen and William P. Schnatter, conceived the M-19. It took just seven and a half months to move from the drawing board to a workable prototype for the M-19. Much of this effort was performed—in the mid-1960s—at the Naval Ordnance Station in Louisville, Kentucky, "to meet an urgent U.S. Navy requirement in Vietnam." Eight hundred of the Louisville-produced M-19s were sent to Vietnam, but reliability problems (not firing, safety questions, and so on) interfered. Testing facilities were moved to Dahlgren, Virginia, where Colonel Chinn supervised modifications. Six hundred M-19s were sent to Vietnam from Dahlgren, many of which fell into the hands of the Viet Cong when they stormed the capital of Saigon in 1975.[30]

Some of Chinn's military colleagues referred to the M-19 grenade launcher as a cross between an automatic rifle and a cannon. The M-19 could "take out" a treetop (a refuge for enemy snipers) with one shot. It provided an infantry unit with "artillery capability."[31] Effectively mobile, the M-19 is 43 inches long and weighs 72.5 pounds, with the tripod on which it must be mounted weighing another 44 pounds. The M-19 can shoot 300 pounds of "high explosive projectiles" a minute at ranges of one and a half miles (some 1,500 meters), while firing 325–375 rounds per minute, known as the "cyclic" rate.[32] Realistically, rates of fire for the M-19 are 60 rounds per minute in short bursts and 40 rounds a minute in sustained fire. Otherwise, the barrel would "burn out" in a couple of minutes.[33] In some opinions, though, the M-19 will *not* burn out at the higher rate. Many machine guns are equipped with a spare barrel to avoid overheating the mechanism. The M-19 "was purposely designed so that the barrel does not need to be swapped out. . . . Usually a few rounds are fired and then the gunner pauses to see where they strike and then redirects the next few rounds, and so on and on."[34] Whatever its capacities, one thing is for certain about the M-19: it is a formidable military weapon.

A minimum crew consisted of a gunner, an ammunition carrier, and, especially out in the field, additional personnel to carry the tripod and other equipment needed for a smooth and effective operation. The M-19 is also designed for use on helicopters and river boats, always with tripod mounts, and on land-based vehicles such as Humvees and trucks, ranging all the way from pickups to 4x4s.[35] It seems, indeed, to be a quite versatile weapon of war.

After introducing the Mark-19 prototype (the "official" title is "Model"; "Mark" is the phonetic for "M") and with the end of the Korean War in 1953, Colonel Chinn, like so many of his Marine Corps and other military compadres, winged his way home, in his case to Louisville and then Harrodsburg. He cleared out enough of the cave (the old Chinn Cave House had closed in 1938) to set up a firing range—which he rented to the U.S. Navy for $1 a year—primarily to test the capabilities of the M-19. He wrote his *Machine Gun* books from his "office" in Harrodsburg—a dozen miles away—during this time, and went to the firing range at every opportunity, with a resultant near loss of hearing in both ears. Also, because he was around so many weapons firing off, dust and tiny particles of brass became embedded in his face; he had to undergo cosmetic surgery in the years ahead, including even small pieces of skin grafts.[36]

In his "off time," if he ever got any, he relished going to his and Cotton's house situated on the high cliff above the Chinn Cave House. On one occasion a couple of military officers visited him at home where, from his front porch, they "plinked" all over the place. (As noted earlier, plinking is shooting a rifle at anything not sentient.) Their targets were "real" ones for "real" marksmen, especially if the wind were blowing and the shooter had to follow the target. But then one of the military fellows began to ask persistently about a machine gun. The colonel smilingly took him down the hill to an assistant whom Chinn ordered to fire the weapon. The military visitors were afraid of being arrested by the county police, but Chinn assured them that the local gendarmes would leave him alone. Hearing the noise, some neighbors might possibly come out and ask the colonel if he needed any help (with what they never said). "Are the Yankees getting closer?" the colonel usually asked.[37] And there the matter ended; Colonel Chinn could fire as many guns as he wished in the palisades area of Central Kentucky. It was not unknown for him to sit on his porch overlooking the Kentucky River and shoot the M-19 at other outcroppings of the palisades. To anyone who argued with him about the noise and damage, he replied that his ownership of the place gave him the right to shoot it.

But before all the highly workable attributes of the M-19 prototype could be fully employed, Chinn and his crew had to work on this weapon for several months during the 1950s while Chinn wrote his voluminous account of machine guns. There was some urgency here in perfecting the M-19 because many military and civilian minds suspected that the major area of American future combat would be in Vietnam. At the battle of Dien Bien Phu (March 13–May 8, 1954), it was obvious that the French were out and it would not be long before the Americans were in. U.S. military and civilian officials predicted that the next war, in all likelihood, would be in Southeast Asia, and that if they were correct in this supposition, the M-19 would certainly be an effective weapon in the jungles.

The first problem with the M-19 was the so-called o-ring. The o-ring was installed at the breach end of the barrel, acting as both a gasket to seal the gases around the projectile (grenade) and a cam to bring the projectile into alignment with the barrel. This ensured that the projectile maintained a straight path throughout the firing cycle. When the operator fired the gun, however, the o-ring frequently obstructed the barrel (which was larger than the machine gun from which this armament was adapted; the projectile was a grenade, not a smaller bullet). This created a distortion of the ammo's path, leading to inaccuracy. This flaw led to what became known as the "curved rail" in the final product, eliminating the o-ring. In earlier versions of the M-19, the loading and ejection of the grenades sometimes created a jam when the operator tried to fire it. The curved rail controlled the "cycling" of the M-19. As one military weapons scholar put it: "As the bolt comes forward a live round is chambered and fired, while a fresh round is picked up. Some of the energy of the recoil is used to force the bolt backwards. The curved rail pushes the fresh round down into firing position and this, in turn, ejects the spent casing out the bottom of the gun. If the trigger is 'off' the cycle is over. If the trigger is on, the weapon will continue to cycle."[38] The curved rail eliminates the extractor and ejector, making it impossible to stub a round. The spent projectiles drop off onto a container below the bolt or onto the ground beneath, thus preventing, as with the o-ring, a traffic jam.[39]

The M-19 was used primarily in foreign conflicts; domestically, its threatened use in the 1950s and early 1960s alleviated some civil rights disturbances. Baton Rouge, Louisiana, was one place where just the possibility of a Mark-19 being used against demonstrators quelled the situation.[40]

The last major military instrument invented by Colonel Chinn was not a weapon at all; it was a "flameout eliminator" to prevent flameouts on high-altitude test planes. A flameout happens when the jet engine stalls owing to the flame going out, usually because it is starved for air. The high pressure of the weapon discharging close to the intake area of the jet engine probably starves the jet engine of air, thus causing a flameout. Chinn's invention deflected the gases of the weapon away from the intake area so that the flame would not go out.[41] Chinn's solution for this problem won him high praise from one of the military's chief test pilots, Colonel John Glenn.

In the mid-1950s, after the Korean conflict had become an uneasy armistice between north and south, Colonel Glenn test-piloted the new high-altitude naval plane, the F7U, leveling out at twenty-nine thousand feet, with an absolute height of thirty-nine thousand.[42] The guns were in the vicinity of the jet engine air scoop, which apparently contributed to engine stall or sometimes complete shutdown at high altitude during gun firing.[43] Half a dozen naval engineers worked on the problem to no avail; finally, George M. Chinn invented the aircraft machine gun muzzle blast diffuser, "which has been of particular value in the prevention of jet engine flame out in a late model Navy fighter aircraft."[44]

As is the case with the F7U-Cutlass, when a cannon is mounted with its muzzle near the intake area, a special danger can develop when the cannon is fired at high altitude. When the cannon is fired, a cloud of hot gases follows the projectile out of the barrel. This cloud momentarily can block air flow into the jet engine and cause a flameout.

The diffuser was made of stainless steel and resembled the "cuts" of a sawed-off shotgun. These were slots or louvers that "redirect the gas and reduce recoil as the shot emerges from the barrel." After attaching the flameout onto the cannon barrel, Glenn and Chinn taxied the F7U to a ground firing range and "with engines running," fired the guns. "The instruments showed greatly reduced" possibilities of flameout.[45]

Then Colonel Glenn took the plane up in the air. At twenty-eight thousand feet he fired the guns with no evidence of any malfunctions, flameout or otherwise. The problem with the guns had usually started at that altitude; that there was no difficulty this time gratified both Glenn and Chinn. Glenn took the plane up thirty-nine thousand feet, its maximum altitude, and for two flights in a row, it seemed that the diffuser had done its job—certainly a cause for celebration. Then on the third flight at maximum altitude, Glenn reported, "There was a godawful racket, with extreme vibration, and it looked as though every emergency light in the cockpit was lit up."[46]

Colonel Glenn meticulously maneuvered the F7U back to the runway at Patuxent, where investigations began immediately. Colonels Chinn and Glenn, with their respective crews, delved into the vibration problem. They found that the repeated gunfire, both on the ground and in airborne experiments, had weakened the metal in the diffuser, which had flown off and "literally stripped the engine."[47] Once the firing operations were fixed, the F7U was back in service, at least for a while, only to be superseded by the F8U-Crusader.

Colonel Glenn effusively thanked Colonel Chinn for the latter's work on the F7U: "Colonel George M. Chinn was a career Marine stationed at the Naval Ordnance Depot at nearby [from Patuxent] Dahlgren, Virginia. He was one of the world's leading armaments experts and had developed a number of weapons."[48] Chinn also basked in the compliments of General Pat Patrick, who said that the "flame out eliminator" allowed the mounting of weapons "in close proximity to the jet planes air intake scoops" and was, as a result, invaluable to the operation and safety of both the plane and the pilot. Chinn served on a "fix-quick team," a group that handled, on a worldwide basis, "all automatic weapons' malfunctions whether afloat or ashore." The general closed his remarks about Colonel George Morgan Chinn by saying: "His profound knowledge of automatic firing mechanisms with his high professional skill in adapting them to specific combat needs has made it possible for United States forces to bring superior fire power against the enemy."[49] For once, the colonel did not know what to say in response to someone else's remarks.

The Fleet Marine Force, headquartered in San Francisco, sent commendations to Chinn: "His ingenuity, technical skill, and determination" were instrumental in safeguarding not only Colonel Glenn but Colonel "Chuck" Yeager and literally dozens of other test pilots as well with the installation of the flameout eliminator. This "vital project" spearheaded by Chinn "resulted in a notable contribution to the efficiency of our arms. His conduct throughout was in keeping with the highest traditions of the United States Naval Services."[50]

How did George Morgan Chinn perform all these duties? He did not get information and instruction from any engineering textbooks or, for that matter, any other type of book. He had no college degree in engineering. (He had majored at Centre in journalism and phys. ed., with a distinct emphasis on the latter.) Well beyond the improvements to the Browning .50 caliber machine gun, the M-19 automatic grenade launcher, and the flameout eliminator, he worked on literally hundreds of other weapons, some malfunctioning so badly that they would not even fire. He and his crew got them up and running again, much to the good fortune of U.S. military efforts. In a way— although he would probably object to this description—Chinn was a primitive genius, that is, one of a kind, in reference to big guns used by the military. This put him into the company of such notables as Grandma Moses, who never took a painting lesson in her life; Zane Grey, author of nearly a hundred western novels, who disdained writing classes; and Billy Vaughn from Glasgow, Kentucky, who could play five different musical instruments without ever having taken a music class, let alone being able to read notes.[51] George Morgan Chinn should be counted in this group; like other primitive geniuses, his talent was pure instinct. Are people born with such instinct or do they acquire it? Perhaps his genius dated back to the time when C. W. Longmire in Frankfort let George take a Gatling machine gun apart and then figure out for himself how to put it back together. He finally learned how to reassemble this weapon, and it certainly was a major accomplishment for a five-year-old lad. The years ahead only added to the expertise he gained through his fascination with firearms, especially big ones needed by the government in times of war.

It was trial and error almost all the way with George Chinn and guns. He was not keen on taking parts away from big guns; on the contrary, he usually wanted to *add* mechanisms to refine the qualities of the gun on which he and his crew were working. If the additional parts did not fix the problem, he'd take them off and try something else. Pragmatism, therefore, was one of the hallmarks of Colonel George Morgan Chinn. Now and then one of his colleagues would ask why he did not always write down his plan of attack toward a malfunctioning weapon, instead of looking at it, feeling it, listening to it, trying to fire it. He almost always replied, "I've never had a malfunction on paper." Invariably, however, one or another of his crew would misread instructions. Much earlier than the country tune by the same name, George always liked to say, "You can make something foolproof, but not damn fool proof."[52]

This commonsense approach usually rattled a few nerves, but his crews always took him seriously and performed in professional ways. His innovations helped considerably to defeat the Axis powers. Chinn and his colleagues continued, well after World War II and Korea, to receive high acclamations from top-ranking military and civilian officials.

Chinn thought he had taken military retirement after Korea, and he headed back to Harrodsburg for a "normal" lifestyle. He had been in the Marines during World War II and Korea, and this tenure added up to around twenty years. He would serve later in the Vietnam War; seemingly, he could not get away from the military. George Morgan Chinn did not complain about the situation. He *liked* being a colonel in the U.S. Marine Corps, and he made the most of every minute he served in it. In the late 1950s and early 1960s the colonel chafed for something to do, bored by the "dull" peacetime period. Again with Happy Chandler's help, he found it.

5

∾

"Whose History Is It, Anyway?"

As the 1950s progressed, the colonel constantly grumbled about having nothing to do, even considering himself unemployed. As he complained of ennui, he continued to work on the M-19, wrote (with a secretarial staff) the last volumes of his masterpiece on the machine gun, visited Frankfort as often as he could (where his old buddy Happy Chandler kept saying he'd find something for him to do), and became interested in local subjects such as historical preservation, especially of nearby Fort Harrod, civic clubs, and even the humane society. On top of all these "nondoings," he was frequently invited to speak to numerous organizations (usually about fealty to one's country) in Harrodsburg and surrounding towns and villages. He was also reformulating a few plans for the resurgence of the Cave House, which had closed in 1938. He had a difficult time convincing other people that he really didn't have anything to do.

One thing he had to do was help his old friend Bud Dedman publicize the area and its attractions (Dedman's family owned the Beaumont Inn). Frequently Chinn and Dedman tied two houseboats together and went up the Kentucky River with "several travel editors and writers" on board, the reasons for which were obvious.[1] Often, these tied-together boats carried a full complement of booze and music, accompanied by singers, dancers, and vaudeville-like entertainers. They were not quite the same as the great "showboats" that paddled down the Mississippi and other major water thoroughfares in an earlier generation, but they were certainly big enough to catch the townsfolks' attention and encourage them to attend performances.

Chinn never turned down a speaking invitation from civic organizations: Rotary, Civitan, Lions, Jay-Cees, churches, and schools. His speaking topics generally had to do with love of country, World War II, and Harrodsburg (which, he ardently believed, was the oldest settlement in the Commonwealth, predating Boonesborough. He was ultimately involved in a pretty rowdy series of events in regard to this question). He also loved speaking to local Marine and naval groups each November 10 on the birthday of the Marine Corps (which had been commissioned in 1775). He rarely missed giving a speech to the Leathernecks in the Bluegrass area on that date between the late 1950s and mid-1980s. He had relished just about each moment of his twenty-six years in the military and fully enjoyed imparting his knowledge and experiences to younger audiences. He spoke to Marines, retired and on active duty, in all likelihood, at Beaumont Inn. He lightened things up a bit when he told his audience that he was "happy to be here tonight," adding, "But anybody at my age should be glad to be anywhere."[2]

The passage of time often reminded him of his marriage to Cotton. Though audiences rarely asked for it, he gave them some marital advice anyway. When a quarrel between spouses reached a reading of 9.7 on the Richter scale, the one who was losing should leave the house and walk around "until they had cooled down" enough for normal conversation. Such a practice, he maintained, had greatly contributed to the tranquility of his own household, "but it is also responsible for my good health due to the thousands of miles I have walked out in the good fresh air." He counseled that a man should keep "suppressed rage" to himself before going to sleep at night: there was no point in getting in "little digs" at the wife "that will send her airborne."

The strategy worked for Colonel Chinn—he usually woke up each day calmly and serenely. "But," he deadpanned, "I did find it a bit hard to go without a wink of sleep for as much as 4 or 5 weeks at a time."[3] Audiences loved these moments of frivolity and returned time and time again to hear him speak, even though he sometimes used the same lines.

He did get serious in these Marine Corps speeches and other discourses. He said he did not have the "emotional stability" to look back on his personal experiences in World War II and the things he had seen, but there was one event that he solemnly related to his audiences. Returning from Japan just as the war ended, Chinn's plane made a fuel stop on a tiny Pacific island that was a staging area for the bodies of dead Marines before they were moved to Arlington National Cemetery or family burial grounds for final rites. "Wishing to pay my final respects I was given a chart showing location of plots and an alphabetical listing of the dead. As I walked through the seemingly endless rows in silent bivouac, I noticed many familiar names. I will not attempt to describe the awesomeness of the peaceful atmosphere and the feeling I had been selected by the lottery of fate to have revealed to me the formula for the mysterious glue called 'esprit de corps' that binds all Marines together."[4]

Many of his speeches rebutted the sentiment of the great Roman poet Horace that "it is sweet and fitting to die for one's country." Chinn told a Memorial Day audience in Harrodsburg on May 30, 1956, that Horace's reasoning was askew. "The tragedy of such a philosophy," Chinn exclaimed, "is that it takes the flowers and best of our youth who otherwise could be strong and useful citizens." War "is so useless." Two world wars and Korea had been fought in his own lifetime, all proposing to bring peace and prosperity to the citizens of the world. But that promise had not been fulfilled, and moreover, Vietnam waited in the wings.[5] His theme in this speech, addressed to the mothers of military sons and daughters throughout the country, was that the military does not cause wars, civilian politicians do—and then the military has to fight the enemies in the ensuing conflict. Surely, there must be a better way than war for the world to manage its differences.

In Mercer and some adjacent counties, George Morgan Chinn was considered a great American hero. He had actively served in two different conflicts, been highly decorated and honored for his efforts and, most important, "had brought the boys home." Some even seemed to believe that Chinn had single-handedly defeated the Japanese in the Pacific. Ironically, all this adulation came in the middle of Chinn's life, when he thought he had accomplished little or nothing and had led an unexemplary life. Perhaps he was depressed; he was idle, he thought, not involved in interesting, exciting, and important events.

None of this was self-pity; his ego would not allow that. If his "noninvolvement" with society bothered him so much now, how harsh would be the restrictions of total retirement? No, retirement was out of the question. It was not what he considered to be "advancing age" (he was fifty-three years old in 1955, and thus eligible for a military retirement plan). He often told questioners who asked about retirement (which usually put him into a bellicose mood), "I'm too old to retire." If he retired, things would get too "peaceful" and therefore become "pretty boresome."[6] He knew he'd miss the camaraderie of the Marine Corps, the challenges of supplying one weapon after another to the fighting men and air aces, especially those in the Pacific arena. And, frankly, he'd miss being the "star attraction" of Harrodsburg and whenever he came home on furlough.

About fifty miles north of Harrodsburg on State Highway 27 is Frankfort, Kentucky, capital of the Commonwealth, a place of great historical treasures and home of innumerable innuendoes, "conspiracies," gossips, and endless rumors. State government workers there long ago had dubbed it, like the State Department in Washington, DC, "Foggy Bottom." Colonel Chinn had once said, "If you don't have a certain amount of good, healthy controversy, you'd better be worried."[7] He had no idea at the time what kind of self-fulfilling prediction he had uttered. He had few clues that the next several years of his life would be intricately bound up in the policies and actions of one of the state's agencies, the Kentucky Historical Society, and with several individuals connected with it who thought they owned it.

In the late 1950s the Kentucky Historical Society, founded in 1836 and headquartered in Frankfort, operated on a budget of about $321,000 in taxpayers' money, with a membership of some nine thousand.[8] Most Kentuckians had never even heard of KHS, but many of those who had regarded it as a stronghold of Frankfort's Old Guards, also known as the Blue Bloods. The majority of KHS members at that time came from this group; they were in no way inclined to change either the philosophies or the procedures of the historical society and its genealogical affiliates. They evinced little interest in recruiting members from parts of the state other than Frankfort and surrounding Bluegrass areas. But some academic infusion into the organization, many "Young Turks" believed, was vital to its continuance. Colonel Chinn and Dr. Thomas D. Clark, head of the department of history at the University of Kentucky, were in agreement on this matter. Clark spoke of "haughty aristocrats" running the society; when he mentioned Eastern Kentucky to a secretary of the KHS, she disdainfully replied that she didn't consider that area "a part" of the state.

A vacancy occurred in KHS in 1959. Charles Hinds, who had been interim editor of the *Register* since Bayliss Hardin's death in an auto accident in 1956, wanted Dr. Clark to assume the presidency. The Old Guard, led by director Willard Rouse Jillson, would not countenance Clark in the position. The reasons seem to have been differences of opinion in how to run KHS (Dr. Clark was among those who hoped that ultimately KHS would represent people all through the Commonwealth, not just a select few), and some discord between Jillson and Clark on matters of state government. The board of directors met to elect a new director, Clark or Jillson. Each member had to be physically present at the meeting; voting by proxy was not allowed.[9] Clark lost the election by one vote. In a temper, Clark returned to Lexington, and a short time later, dissatisfied with Jillson in office, Hinds himself left the KHS.

Although KHS continued to laud Dr. Clark as the state's leading historian, Clark would have nothing to do with the organization over the next several years, going out of his way "to run it down" at every opportunity. Ultimately, the *Register* editorship fell to Georgetown professor James Klotter, who had been a PhD student with Professor Holman Hamilton, a highly respected member of the UK faculty. Hamilton and Clark were close friends, and when Klotter began his editorship at KHS, Dr. Clark began to make periodic visits to the society. He had heard that a history center was in the makings; perhaps he could influence the planning and building of this much-needed repository of Kentucky history. A young historian at KHS, Russell Harris, remembers one of Dr. Clark's visits: "He surveyed the pile of books, papers, tapes, and notes littering my desk, and said in his most icy tone, 'Young man, that's the way a desk should look.' I breathed again."[10]

Another person who made frequent trips to KHS was Colonel George Morgan Chinn. Ever since his return from Korea in the mid-1950s, he had been absorbed—more than at any previous times in his life—with Kentucky history, especially that involving Harrodsburg and the Big Settlement in Bluegrass country. He was involved in KHS activities from 1956 onward; he saw a comparison between cataloguing guns and ammunition and using the same process with books, periodicals, and magazines. Each had to be identified as to genre and sort, and assigned an identifying number. He and many of his colleagues, therefore, strongly believed he was more than competent to offer his services to KHS, although his field of expertise was military weaponry.

In late 1958, the director of the Marine museums, Lieutenant Colonel John H. Magruder III, requested that Chinn be brought to active duty for thirty days to travel around the country collecting military weapons: small arms, big rifles (automatic and otherwise), machine guns, and cannons, many of which were now superannuated, some still in use, and a few projected for the future. "It is requested," he said in a letter that went out to all Marine posts in the country, "that Colonel Chinn be afforded every consideration in connection with the solicitation of such material."[11] Magruder wanted Chinn headquartered at Quantico, in Virginia, to work directly for the museum. His duties, which included numerous visits to and studies of other Marine museums in the country, considerably helped the staff to exhibit relics and artifacts correctly. It also helped Chinn develop a "how-to" sense in reference to nomenclature, classification, and sorting. Many civilians and military individuals believed that the atomic bomb superseded all previous weapons. Chinn thought this too arbitrary: Americans "cannot afford to neglect the finest possible weapons of every type."[12] Nuclear weapons must develop *alongside* conventional weaponry. A study of the past and its impact on the present was just as necessary in ordnance as in other aspects of military activities. Once more, Chinn and his friends and colleagues argued that these experiences with the museums added greatly to his capabilities to direct such an organization as the Kentucky Historical Society.

One big question in the whole Clark-Chinn affair was "Where does Governor Chandler stand on these matters?" Did he exercise an influence on who occupied positions at KHS, including that of director? The Chandler and Chinn families were still as close as ever. After one of Chinn's "discharges" from the Marines (he was constantly being recalled), he visited Governor Chandler in Frankfort, pleading with Happy to find him a job. He was as sure on this occasion as he had been when seeking Chandler's help to get into the military: he actually took his suitcase with him. Chinn told interviewer William Ellis, "When I went into the Marine Corps, Happy Chandler was governor and when I got out, he was still governor. . . . I had my suitcase out in his office . . . and the trunk was being sent over to the mansion and I'd live with him till he got me a job."[13] Again, as before the war, Chandler feared that Chinn would eat him out of house and home. Did he thus agree to help Chinn become head of the Kentucky Historical Society? As governor, Chandler was ex officio chancellor of KHS, and his influence was paramount. It is entirely within the realm of possibility that Governor Chandler made a few telephone calls on behalf of Colonel Chinn in reference to KHS. Whatever the case, Chinn served as director from 1959 to 1973 and then as deputy director for a few years afterward.

During Chinn's tenure he and Clark had a standoffish relationship. Governor Chandler considered Dr. Clark the most notable professional historian of Kentucky history in the Commonwealth. He regarded Colonel Chinn as the most highly esteemed ordnance man and as a personal friend. But one was a professional, the other an amateur, and this difference brought about an unending breach between the two. No matter how many times Chinn and the "professionals" (including Dr. Clark) agreed with one another, among the latter there was always silent innuendo and implied questions: "Who are you?" "Where is your degree?"

Clark related a story to interviewer William Marshall about a time he and his wife, Beth, stopped for breakfast in Ripley on their way to Columbus, Ohio. Beth remarked, "It's a strange thing. There is a man over there just looking daggers at you. Do you know who that man is?" Clark looked and said, "Yes ma'am, I know who that man is. I know why he's looking daggers at me, too. That's old George Chinn." The tension between these two men was palpable. Clark said, "In the case of George Chinn, I incurred his anger forever more, and intentionally and militantly so."[14] In reference to intentions, therefore, it seems clear that Dr. Clark did his part to keep the flames alive.

Both men had healthy egos, and Chinn bore the extra load of a persecution complex. Some observers swore that Chinn once referred to Clark as a "carpetbagger" from the South (he was born in Mississippi) whereas he, Chinn, was the "genuine product," that is, from Kentucky and thus more a representative of "local" or "public history" than most "professional" historians, and as capable as Clark to speak and write about it. He told a reporter that "I'll get into it with any PhD. They are boresome but 'correctly' boresome."[15] He added fuel to the fire by asserting to another reporter that "too many histories are written by PhDs and then edited so that it takes another PhD to understand it."[16] Chinn denied, however, that he had called Clark a carpetbagger, and Clark, for his part, did say something positive about Chinn every now and then. These were, however, rare occasions.

Early in his directorship, Chinn called a meeting of the KHS board of directors. The first thing that aggravated Dr. Clark at this gathering was the long table at which they sat, the kind used at military briefings by commanding officers wanting to impart orders and information to underlings; this played into fears that Chinn planned to turn KHS into a military arms museum. "I didn't," Chinn said later, "and I don't intend to." According to Clark, Chinn "explained how he was going to run the Archives and how he was going to take over." Clark rejected the idea of joining the archives and society into one agency. He was fearful, he said, that Chinn would turn the archives into a branch of the Daughters of the American Revolution (DAR), in which political fervor would replace sound historical research and writing. "We had quite a hair pulling," reported Clark, "and I won," although he didn't give any information about his victory.[17] Chinn quickly assessed his situation at KHS, deciding that opposition to him as director came essentially from two sides; if the people weren't convinced he planned to turn the organization into a military arms museum, they believed the rumor about the DAR.

Shortly into his tenure, Chinn noticed that a separate state archives had been created, and that supervision of a significant project, the governors' papers, had been transferred to the editorial staff at the University Press of Kentucky in Lexington. Chinn regarded both these actions as unfriendly; they took authority away from the Kentucky Historical Society and handed it over to other agencies. Of course, Chinn blamed Clark for these "wicked" deeds; though no sound evidence ever surfaced to show that his charges were valid, he continued to believe throughout his time at KHS that Dr. Clark was the "Devil Incarnate."[18] Perhaps feelings of paranoia played into Chinn's beliefs: Dr. Clark was the professional historian with a PhD, while there was some question about whether Chinn had even been graduated from Centre College. At any rate, Chinn was the amateur, and he felt that many people at KHS, spearheaded by Dr. Clark, were out to do him grave injustices.

He found, for example, that KHS's "Communique," outlining both the personal and public activities of the various departmental personnel, was distributed without the knowledge or permission of the *Register*'s editor, G. Glenn Clift, or of the KHS director, George M. Chinn. He wrote snidely to Clift, who was also the society's assistant director, that "only in the Kentucky Historical Society, in all of State Government could such an asinine situation exist, where the official voice of an organization is published with the Editor and Director not being aware of what it contains." The "Communique" carried news about acquisitions, manuscripts being critiqued by KHS for possible publication, and sometimes personal news about people who worked at the organization. "I want it distinctly understood," he fumed,

"I will not for one instant tolerate any publications being used as a means of personal promotion for anyone connected with the Society." One of the "Communiques" had quoted Alfred, Lord Tennyson, to which Chinn responded, "If he [Tennyson] must sing, do it somewhere else." One "Communique," he complained, "could not have moved me more deeply had I been exposed to the bilious burping of a love sick pelican." He urged that these "bad manners" not become "habit forming," saying that the photo reproduction staff would not receive payment for services rendered unless the "Communique" bore the distinct approval of "either the editor or director" of KHS.[19]

Charles Manning, editor of the "Communique," defended his practices. Mr. Clift, Manning said, had terminated in 1959 any connection he might have had with the "Communique." Nor had Manning been advised that Colonel Chinn "wished to see each issue of the Communique in advance."[20] One thing the "Communique" matter made clear was that, in addition to his moods and periodic bouts of temperament, Colonel Chinn was a micromanager of the highest degree. This was his method in the Marine Corps, and he continued it during his tenure as KHS director.

And, in many respects, Dr. Clark micromanaged the UK history department. One report claimed that Dr. Clark "had the reputation of frowning on anyone who tried to write about Kentucky without his approval or permission."[21] If this charge were true, it alone would have alienated Colonel Chinn. He was a USMC colonel, used to *giving* orders, not taking them. In fact, he would have relished disobeying.

Colonel Chinn could find plenty of "enemies" to hate; he could also be courteous, even eloquent. Stories abound about the lovable side of his nature. Visiting Joseph Rankin's house one afternoon to discuss "Kentucky Long Rifles," he declined an offer to "take dinner." Rankin's children, however, offered the colonel some homemade Kool-Aid. He drank several Dixie cups of the concoction, pronouncing it "the best he'd ever had." Another time while he was the KHS director, an elderly couple visited for genealogical research. They lost track of time; when they realized that the library had been closed for over an hour, they rushed to the door. There they saw a portly man sitting in a chair patiently waiting for them to complete their business. The two were greatly embarrassed, but Chinn put them at ease, and they left the building in an excellent mood. In his eighties he bought a small, multicolored (mostly yellow) Ford Escort, which always required considerable "huffing and puffing," considering his size, to get into. He told everyone, "When you age and the vision . . . [isn't] . . . as keen . . . [as in times past] . . . you should drive a bright car so people can see you coming."[22]

He amused his friends by taking a large bottle of beer (unfortunately the brand remains unknown), pouring it into an oversized glass of ice, and guzzling down the contents in two or three gulps. "He was the only man I have known . . . who drank beer on the rocks," said one of his friends.[23] He also drank milk from a huge glass filled with ice. When asked which was his favorite bourbon, he reportedly told friends, "I've only done two smart things in my life and both of them were to quit drinking whiskey. That is too big to be one thing." Once, a disagreeable visitor to Chinn's office at KHS created such a fit of violence that Chinn literally threw the man (not Dr. Clark) out of the building. On another occasion, obviously wanting to get rid of a female delegation visiting KHS, he hit upon the expedient of opening a drawer in his desk and pulling out a pair of Governor Goebel's underwear.[24]

When Edna Flowers applied for a job as a secretary with Chinn in his writings on armaments, he asked one question: "How do you spell 'ordnance?'" She got it right and he hired her on the spot.[25] Another of Chinn's secretaries, Susan Blankenship, remembers that Cotton would visit his downtown Harrodsburg office "every now and then." She and Chinn would sit close to each other, "knee to knee," talking mostly about domestic affairs in their big house on the hill. Once, a KHS employee who obviously could not spell came out with "copywright" instead of "copyright" in the *Register*. Editor Glenn Clift was concerned about the matter. Director Chinn, however, chuckled and said, "Well, that'll be fine. We'll just put a little silver dot over that 'W.' It'll be a collector's item."[26]

While Chinn could be dictatorial with employees of the KHS, he was, at least on occasion, solicitous of their feelings. The salaries at KHS were far from the best in the world; they were downright meager, not enough actually to meet an employee's monthly expenses. "You had to love history and genealogy" to work there. Some employees, however, "did muster up enough nerve" to visit the colonel and complain about pay rates. He asked why they needed raises; it turned out that some of the workers were "eligible for government assistance." They got their raises. He became dreadfully irritable every time he learned that important collections had gone to other repositories because the Kentucky state legislature and the governor "could not find the money" to pay for them. The Durrett Collection went to the University of Chicago, while the Draper Papers were shipped off to the University of Wisconsin. Both these collections had significant files of information about Kentucky, and both were now stored in other states, much to the dismay of both Colonel Chinn and Dr. Clark as well as many other individuals with KHS connections.[27] Clark and Chinn may have had their differences, but they wholeheartedly agreed with one another about keeping Kentucky historical materials in Kentucky.

Even so, the two were at odds over the very definition of history, inflaming their quarrel. The June 1989 issue of the *American Historical Review* discusses in depth the nature of history. Only relatively recently (primarily in the twentieth century) has history become a separate academic discipline. In the past, history was part of other departments—political science, rhetoric and speech, and philosophy.[28] It has been closely affiliated with subjects such as sociology, folklore, psychology, and others. It may be argued, therefore, that history is not an entity of and by itself but, like geography and economics, must be pursued in conjunction with some other subject. No wonder, then, that the question arises: "Who owns history?" Is it the gifted amateur or the professional academic? Or could it be both? And, of course, how about the readers—how should they react to this question? Should they be given any kind of consideration in reference to subjects and interests? Both professionals and amateurs vote "yes" on this question.

A Chinn admirer stated forthrightly that Chinn was "a great believer of 'fact before fiction'" in historical matters, "trying his best to separate the two, and to set straight Kentucky lore and legend which had no basis factually or proof-wise."[29] Dr. Clark, however, continued to believe that Chinn *added* to legend and fiction on more than one occasion. Another historian disagreed with Chinn's viewpoint that the lay historian could match the professional one, saying that it was how materials were *handled* that created the greatest difference between the amateur and the professional. "A trained historian is more likely to dig deeper and know how to interpret the material and its environment. A trained historian follows his notes to their logical conclusions; an amateur tends to agree with material only if it agrees with his biases."[30]

Another history professor, Robert Norrell of the University of Tennessee, Knoxville, claims that "it is clear to me that the academy hardly owns history. Far more books are sold by journalists writing history than by PhDs writing it." Chinn would have been pleased with this statement, Dr. Clark chagrined. "The public," continued Norrell, "has a keen interest in historical subjects . . . but does not rely on academic historians." Some history departments are turning to "public" history to train their graduates for jobs in museums and libraries rather than the classroom.[31]

One professional historian argued that the "academic historian" is a "newcomer to an old discipline." The amateur, on the other hand, has been around for many, many years. "These amateur historians chronicled history on which later generations of professionals based much of their work."[32] A few historians, professional and amateur, accept this point of view and try to lessen the differences that have arisen in researching and teaching history. "History does not 'belong' to one particular group of people, be they scholars or laymen. History is the possession of those who find it interesting and enjoy exploring its world."[33]

There can be more than one legitimate interpretation of a single event, as long as that interpretation is honest. History "is a story we tell," said one writer. "The story changes, new elements are added . . . myths invented . . . facts debunked," and history either confirms or denies them. "History is a process of imposing order on a chaotic process."[34] While both Chinn and Clark would probably have accepted this approach to the study of history, they still stood in opposition to each other, sometimes with hostile and far-reaching results. Dr. Clark was always interested in the history of the entire state of Kentucky; Colonel Chinn tended to concentrate on Harrodsburg, Mercer County, and the Bluegrass regions. Chinn's closeness to these areas caused Dr. Clark to remark that "professional historians have to stay away from even the smell of boosterism."[35] Chinn got the message and became even more obstreperous toward Dr. Clark than before.

Perhaps there was a touch of inevitability in Kentucky having two major spokesmen for its history, one formal (Dr. Clark) and one informal (Colonel Chinn).[36] Even geographically, Kentucky is a "compromise," with divided loyalties between North and South. Dr. Clark reflected much of the former and Colonel Chinn the latter. Chinn gave speeches of praise over the graves of Confederate dead in Harrodsburg and surrounding cemeteries. As president of KHS, he'd go after the leaders of the Pennsylvania Historical Society over such matters as long rifles, Daniel Boone, the "creation" of bourbon, and other choice differences between the two commonwealths. Some of his remarks about these arguments were mere hyperbole, others not.

Chinn received a lot of letters at his home address on Main Street in Harrodsburg. He usually glanced at them quickly and then discarded them as being of no interest. One morning in 1961, however, just as he was beginning to make inroads at KHS in reference to organization, projects, and acceptance, he looked at an envelope and knew before he opened it what it contained. As he suspected, it started out, "Greetings." George Morgan Chinn was once again going to be on active duty in the U.S. Marine Corps for service primarily in the Marine Museum in Quantico, Virginia, and in Vietnam. He was fifty-nine years old.[37]

He was not caught off guard completely; in fact, as a reservist, he had actually been hoping for a return to active duty—out of patriotism for one thing, but also to complete enough years for a comfortable retirement as a USMC veteran of three wars, World War II, Korea, and Vietnam (four, if one adds World War I, in which he was slated to serve when the war ended). His friend Magruder told him about a possible "general mobilization" to face the threat in Vietnam: "You can be one of the first to be called to duty," to make sure the armaments were working. "In fact," Magruder continued, in not all that comfortable a way, "you would not even be spared from your deathbed."[38] The phrase "Get Chinn," apparently, was still a very valid demand.

As far as payments for his duty at Quantico, Chinn would receive pay "providing the kitty so permits." Magruder explained in pragmatic terms, "If funds are not available," Chinn must remember that "the government is not above free-loading and will welcome you to duty at no expense to the government."[39]

Chinn jumped at the chance to stay in the military; on the whole, he loved it. He did think—at least, from time to time—about moving back to Mercer County and living with Cotton in their hilltop home in happy retirement. He was too much a doer, however, to dwell on this point very long. He was Marine, and that's all there was to it.

As it turned out, he was not generally a part of any mobilizations at this time heading for the Far East, but he did go to Vietnam. His experience in the war helped him to develop a decidedly conservative mind about the contentious issue of Vietnam. It was shameful how some of the returning GIs were treated by the American public. Were there any "Flower Children" (such as in San Francisco at the time) in and around Harrodsburg? asked an interviewer. Or any "Peace Makers?" "No," the colonel replied abruptly. "We would have killed the SOBs." He continued, "Peace and Love, and all this stuff just wasn't accepted around here."[40]

When his tenure was up at Quantico and Vietnam, Chinn ventured back to Harrodsburg, where he was welcomed as a hero, and to Frankfort, where he was largely regarded as an adversary in reference to the Commonwealth's history. Even while he continued to serve as director of KHS, he had other projects in development. The second most significant thing on his mind (the KHS was first) was the cave out on Highway 68. He and Cotton had made names for themselves back in the 1920s and 1930s for Chinn's Cave House. Now he proposed to open it again for treasures that he may have missed in the past (for all he knew, there might be magnificent stalactites and other wonders that eager tourists would love to pay to see), and for the construction of a legitimate shooting range. The KHS and restoration of the Cave occupied him for the next several years of his life.

Braxton Hall school. George Chinn is standing on the right, wearing a straw hat.

The Chinn home overlooking the Kentucky River at Brooklyn, circa 1930.

George Chinn, circa 1920.

George Chinn (*left*) and friend at Mundy's Landing, circa early 1920.

Chinn's Cave House at Brooklyn, Mercer County, circa 1930.

Chinn's Cave House, circa 1930.

Chinn's Cave House, circa 1930.

Quantico, Virginia, graduation ceremonies, circa 1942.

Oldsmobile Aircraft Armament Training School, 1943.

Colonel Chinn (*left*) in Korea.

Colonel Chinn in uniform.

Chinn earned a service medal for his service in Korea.

Chinn worked to enhance the performance of large weapons.

Colonel Chinn, USMC.

Commanding officer Captain R. L. Sattler presents to Colonel Chinn the Navy Meritorious Public Service Citation. At left is Chinn's daughter, Mrs. Howard Howells, and next to him is his wife, Haldon.

6

∾

Back to the Cave

Despite world travels, military armament duties, and KHS responsibilities, George Morgan Chinn never forgot the Cave House on Highway 68 that he and Cotton had lovingly run for some years. They closed it in 1938 because profits were dwindling and because George sensed strongly that his military career loomed in the short distance. Around that time, for $1 a year, Chinn leased the front parts of the cave to the U.S. Navy for ordnance experiments. While on active duty, he frequently thought about the cave, considering what he and Cotton might do with it once he returned to civilian life.

It was by happy coincidence, then, that one day in early January 1966, while Chinn worked at his KHS office in Frankfort, two well-known area spelunkers came to see him. Russell Hatter, a radio announcer for Frankfort's WFKY, and his friend Paul Clifford had explored many Bluegrass caves. One day, after an unsuccessful probe into a cave near Georgetown, Hatter returned to the radio station and told janitor John Adams of his and Clifford's "unproductive" day. Adams asked Hatter if he had ever heard of a big cave on Highway 68 that belonged to George Morgan Chinn.

Learning that George and Cotton Chinn did indeed own Chinn's Cave House, Hatter and Clifford headed straightway for Chinn's office. "As soon as I mentioned the cave, he got all excited—showed me some pictures, too."[1] Chinn's youthful dreams of finding something in the back parts of the cave—like stalactites and stalagmites, and perhaps waterfalls—were quickly rekindled. He had hired people, even while in active service, to "hose out" the cave as much as possible; by 1966, workers had hosed 640 feet of mud and rock out of the cave.[2] On one occasion, well before Hatter and Clifford's involvement, Chinn talked the entire Nicholasville Fire Department into coming to his cave and clearing out the mud and dirt with heavy sprays of water. Chinn had been "on edge" during this operation, fearing that a fire might erupt in Nicholasville while the truck was gone.[3] Hatter, however, remembers Chinn's optimism and belief that "there's something real big back there" in the cave.[4]

Hatter and Clifford were told: "Go right at the 'Y'" in the cave. And then they were to go back until the "mud slopes up to the ceiling." When the two explorers reached that point, they heard water "gurgling" on the other side of the mud, which looked as though it were several feet deep. Hatter also felt a current of air blowing through the mud. The two realized that they could go on if the barrier of accumulated mud were removed. Hatter got to his knees and began to dig. He "grunted" and "crawled" as he and Clifford dug through the tunnel, following the stream of water and air coming toward them. Finally Hatter could not continue. The cave's roof was only two or three inches above the stream and the "mountain of mud" that had amassed over the years. The two spelunkers could go no further, at least on this occasion.[5] They crawled back to the colonel, who waited for them just outside the cave's entrance, where he used to sell gasoline, ham sandwiches, and "needle" beer. The mud had not been there, the colonel said, the last time—just a few years ago—that part of the cave system had been investigated.

Outside, the three men discussed strategies for getting rid of the mud. Nothing by way of finding artifacts or relics could be accomplished until that was done. The first thing on the agenda was to string up lights where there were none (the front parts of the cave had been illuminated years ago) in order to be able to keep a steady watch of the movement and direction of water and air in the cave. They also envisioned an elaborate speaker system allowing communication between those in and those out of the cave. Very quickly Hatter became enamored with work in Chinn's cave, so much so that he even considered resigning from his job at the radio station to devote all his time to this fascinating cave out on Highway 68. Colonel Chinn gave Hatter written authorization to purchase needed supplies for the continuance of the cave's exploration: sockets, bulbs, hip boots, and other small items came to $6.08.[6]

On January 5, 1966, Hatter and Clifford electrically wired up the cave, making it easier to see its curves, contours, and promises. While the two worked in the cave on January 6, it rained all day, and they were gratified to see that none of the rainwater seeped into the cavities. However, Hatter reminded everyone that sometimes it takes twenty-four hours for water to make its way through the earth to any cave systems below. Observation of results here indicated a negative situation, much to the delight of all personnel digging out Colonel Chinn's cave. Hatter and Clifford wanted to dam up the "little spring" in the far recesses of the cave, to direct its flow out of the entrance, taking mud and dirt with it, rather than let it seek a "side" outlet and run-out as was usually the case. "We want this mud washed out of the Main Tunnel," Hatter and Clifford asserted.[7]

A few days later the pair had an opportunity to see the results of their labor when they unplugged the dam they had built. Water rushed out down the main entrance tunnel with "a tremendous force." Colonel Chinn, near the entrance, said he thought it was a tidal wave. He "was frightened for us," Hatter reported. Then, Chinn heard Hatter and Clifford laughing and was relieved. That was the incident, he said later, that convinced him that Hatter and Clifford truly were "professional spelunkers." Anyone else, he argued, would have "cut and run" when the water cascaded from beyond the dam.[8]

Chinn said the water must have rushed over a ledge, for it sounded like falls to him. (This was probably wishful thinking; to find falls or to confirm lack of them in the back parts of the cave was one main reason he had spelunkers there in the first place.) For his part, Hatter could not wait to widen the hole at the dam he and Clifford had built "so I can crawl through it and see what lies beyond." After forty-five minutes of unplugging the dam, the water had receded only about halfway, "so we must have really had it backed up in the rock," Hatter explained. "Discouraged, dirty, and disgusted, we left the mud and sand." They had not been able to stem the tide of the water sluicing off into numerous side areas of the cave and redirect it down the main channel to the outside world. Such accomplishments, they believed, would have to wait for a better day.[9]

And the very next day Russell Hatter was back at the job, hoping that the water had receded enough for him to reenter the interior. He'd crawl back, he said, to the wall of mud. "Once I'm there I'll dig my way through the mud and see what this big water sound is. If possible, I'll dam it up so it will force the mud out."[10]

While Hatter and Clifford explored the inner workings, out in the sunshine, Colonel Chinn and a few family members and friends traversed the earthen and wooded areas above the caves. They estimated that the caves were formed in the Ordovician period (at least a million years back), their formations known as High Bridge and Lexington. Within High Bridge were Camp Nelson limestone and types with other nomenclatures like Dragoon and Tyrone. Lexington represented Curdsville, Logana, Jessamine, Benson, and Perryville. Thickness of the palisades ranged from 15 to 315 feet. Topographical matters were important, for rain and snow could indeed have a significant effect, negative or positive, on the caves below. On November 3, 1968, however, Hatter wrote in his diary: "Colonel Chinn and I have traversed the formations above the cave. Chinn and his son-in-law (Mr. Howells) have been cutting out all the undergrowth. This renewed interest in the cave is good for me." Right after this statement, however, the diary ends abruptly without any farewell, fond or otherwise, and with no explanation why. Perhaps Hatter and Clifford were tired of getting muddy for what increasingly became known in local circles as a "lost cause," though as Chinn traveled from Harrodsburg to Frankfort for his KHS duties, few people actually knew of the activities going on at the cave.[11]

In 2013 Russell Hatter recalled: "The cave venture proved to be a bust. The farther back we went the smaller and narrower the cave became." Nevertheless, just behind a thick wall of mud they could hear water falling in what they assumed was a big room, but Hatter and Clifford never could see it. They did not have the right kind of equipment (automatic drills, for example, and other digging materials), so they left the area on November 3, 1968, and never returned.[12]

It could very well have been that Colonel Chinn had overcommitted himself, taking on too many projects, each of which needed attention, preferably on a daily basis. He was director of the Kentucky Historical Society; writing machine-gun books; relishing a recent award from the secretary of the U.S. Navy, Paul Ignatius; working with General Electric Company's aircraft armament division, visiting its headquarters in Burlington, Vermont, twice a month, consulting with the personnel of the Ordnance Station there; and trying to excavate a cave—all at the same time. He was quickly coming up on his sixty-sixth birthday; he used to tell his friends that he was no "spring chicken" anymore. He felt he needed to slow down. He finally decided to reassess his activities and curtail some of them. But not many.

The letter from Secretary Ignatius to Chinn awarded him with the Navy Meritorious Public Service Citation, the highest honor the U.S. Navy can grant to a civilian.[13] Ignatius notified the commanding officer at the Louisville Ordnance Station so he could arrange an appropriate presentation to Colonel Chinn.

Assuredly, Chinn had won many awards and letters of commendation, dating back to World War II. He cherished a letter written on July 1, 1945, from A. R. Matter of Armament Testing, stating that Chinn was "unqualifiedly the most expert officer on automatic aircraft weapons known to this command. . . . By sheer genius of analysis and inventive power he has been able to discover gun faults and correct them. He is extremely intelligent, and possesses a powerful memory."[14]

The honor from Secretary Ignatius, however, seemed to touch him the most, because it gave him a chance to champion one of the Marines' greatest assets: the service's military museums, especially the one at Quantico, Virginia. Listing himself as director of the Kentucky Historical Society, he wrote a happy letter to General L. F. Chapman, Marine Corps commandant, lauding the help he had received in developing the M-19 and M-20, both of which had progressed beyond the original prototypes at the Louisville Ordnance Station. These two weapons and others were designed to "meet the immediate demands of military operations in Vietnam," especially on riverine operations. The "quick fulfillment" of this project (the M-19 and M-20) would "have been impossible, had it not been for the valuable support received from the Marine Corps Museum."[15] The collections it had amassed from the Korean War (from both friend and foe) made a case against concentrating only on weapons of mass destruction (such as atomic power), arguing that conventional weapons would be of the greatest use in future wars. This supposition turned out to be dreadfully true in Vietnam, Iraq-Kuwait, and Afghanistan.

General Chapman wrote back to Chinn congratulating him for winning the Navy Meritorious Service Award and praising the work of military museums. Too many people, the general exclaimed, thought of museums only as "windows to the past," but they were so much more than that. Many tools that may be stamped today as "no requirement" for military uses will be needed "to meet the demands of tomorrow." The Marine Corps museums, the general assured the colonel, "will continue to serve the needs of research and development as well as those of the historian."[16]

Though he may have been riding high on favorable publicity, Chinn opened up to a reporter that for the past several years he had been under great stress in the Marine Corps. On the surface, one would never have guessed that Colonel Chinn was in any way leading a stressful life. He was jovial and wise-cracking with his comrades in arms and respectful to his commanding officers. But the military schedules Chinn had to meet were grueling. Sometimes he "island hopped" in the Pacific theater, going from one little atoll to another, to make sure the fighting men had reliable, up-to-date armaments.

"I was killing myself eating," Chinn told the reporter. "The more stress under which I was working, the more I ate. And the more I ate, the heavier I became," topping out one time at 330 pounds.[17] He didn't have to be told that he needed to do something about it. He found life at the KHS, even with all the criticisms from academics and others, to be less stressful than maintaining and repairing weapons at numerous posts in the United States and throughout the Pacific. The KHS job gave him a chance to go back to one of his first loves, history. At least, unlike World War II and Korea, it wasn't a "shooting war." He'd facilitated enough actual shooting wars in his life, and now he greatly preferred some rest and leisure. He was only partially successful in this endeavor. But he did finally get his weight down to about 230 pounds and kept it fairly constant during the latter years of his life.

As one might imagine, he was asked many times about his vast knowledge of guns, both small and large. He was so impressive in this category, as well as others, that the *Harrodsburg Her-*

ald wrote an editorial about him. He was asked by a reporter to explain his affinity with arms. "Lowering his eyes and bowing his head he said, 'That's the strange thing about it—I have studied, of course, but thoughts just come to me. I guess I can just say that it is a God-given talent.'" The editorial continued that "since that conversation with the Colonel we have admired the 'Big Man' for his sincere humility and realized that true greatness is always accompanied by a humble spirit. Yes, the 'Big Man' became a 'Bigger Man' that day."[18]

This "humble man" image may have inspired Chinn to offer an invocation, "The Benediction of the Confederacy," for Confederate soldiers buried in the Spring Hill Cemetery in Harrodsburg. Although his grandfather John Pendleton Chinn fought for the South in the Civil War, there is no evidence that George Morgan Chinn was a "confirmed" Confederate. He did like to josh about "Yankees," especially when he talked about Governor Goebel and others from the North. And his viewpoint was always centered on the South. In the 1950s, when he heard someone speak of Ike (as in Eisenhower), he always thought they meant Isaac "Ike" Shelby, the first governor of Kentucky after it had joined the Union.[19]

He truly loved the United States, the country whose uniform he wore for twenty-six years. He said in his speech at the cemetery, "In this friendly ground, far from their homes, for one century, has slept the flower of the Confederacy. They have been our most honored guests throughout these years. Their burial was simple, a few sincere words, a fervent prayer, and flowers still wet from a mother's tears. . . . God, too, must have cried."[20]

An important diversion for Chinn—and he needed one—was the discovery of a couple of firearms in the Kentucky Military Museum in Frankfort. One was a machine gun with a forty-five-degree curved barrel—this was no joke, it was the real thing. The curved barrel was used from 1942 and adapted to the .45 M-3 submachine gun, which fired unexpectedly sometimes, so

fast that GIs began to call it the "grease gun." Colonel Chinn
was called in to the museum (of which he was head, as well as
still deputy director of KHS) and asked to evaluate the grease
gun. He noted that it had all the refinement of an anvil. Despite
his low opinion of the grease gun, however, Chinn said it did
serve a "practical purpose." It was designed for tank commanders who wanted to avoid enemy troops walking up behind their
vehicles and dropping hand grenades or dynamite into the turrets. "It doesn't look as if it would work," Chinn said in wonder-
ment, "but evidently it did very well what it was designed to do."
Soldiers inside the tank simply stuck the weapon out the turret
and pulled the trigger; if enemy combatants were anywhere near
the tank, they'd be annihilated by the grease gun. Chinn was
awestruck looking at this most unusual of weapons. "How that
bullet ever came out of the barrel, I'll never know," he marveled.
"It was quite a piece of engineering."[21]

Then, for a while in 1979, he inadvertently became involved
with what was known as "the glove gun." This was a glove with
a concealed box (containing a high-velocity air pistol) mounted
"on the back of the hand." If the wearer made a fist, it cocked
the plunger. When you "punched somebody in the back, it
discharged!" *Courier-Journal* journalist Byron Crawford wryly
remarked, "No tellin' how many agents were killed or maimed
'cause they forgot to take their gloves off . . . before knocking
on a door."[22] The *Courier-Journal* had barely hit the streets the
next morning when Chinn's telephone began ringing. He finally
answered, and learned that it was someone from Virginia, or
so the person said. It quickly became clear that the caller was
from the Central Intelligence Agency. He told Chinn, somewhat
ominously, "We'll be by to pick up the glove gun." A man who
worked for the telephone company told columnist Crawford
that Chinn's big stone house atop the palisades had been wire-
tapped, and that's how the CIA knew so much about him.[23] The
CIA, without explanation, did get the gun. Many people who

knew the story speculated that the glove gun was an assassination tool; others thought that it was a hoax, merely an attempt to help build the CIA's "espionage" image to the public. Some said it was too close to a James Bond film to be believed.

In 1987 Chinn talked with Professor William Ellis of the history department at Eastern Kentucky University about the aches and pains that usually accompany the aging process, telling the historian that he had been having "throat problems." He thought he'd had an attack of double pneumonia; it turned out to be laryngitis. He told Ellis that "if you ever hear of a murder and it's me" who did the killing, "you'll know that I found the Son-of-a-Bitch that called this 'the golden age.'"[24]

From the early 1960s well into the 1980s, George M. Chinn led alternating lives. He was still involved with the military, primarily now in reserve (but many times in consultation with ongoing ordnance projects) and active in civilian life at the Kentucky Historical Society. He perpetuated the image of a jovial, storytelling (mostly on himself) gentleman to contrast with his secret suffering from stress. Some people said of him that he never let fact stand in the way of a good story. On the other hand, his admirers always credited him with trying to separate fact from fiction. He continued to be a great hero to many in the community, especially children: he seemed to symbolize love of God, family, and country as well as the glory and honor of the Marine Corps. He was not all things to all people, but many individuals with whom he came in contact thought he was. He always had numerous irons in the fire. People and organizations came to Colonel Chinn for advice, counsel, and speeches. He was the personification (except for his perennial weight problems) of someone to look up to, to emulate. His services in both military and civilian life endeared him to the citizens of Harrodsburg, Mercer County, and the Bluegrass areas.

7

_

Action at the Kentucky Historical Society, 1959–1973

Colonel George M. Chinn wore many hats during the 1960s and early 1970s. He was a caveman, hoping to find uncharted caverns out on Highway 68; head of a significant branch of state government, the Kentucky Historical Society; a governmental consultant on military ordnance, especially from a base in Louisville, where he had a direct telephone line; and a much-in-demand public speaker. Taking charge of the KHS in 1959, over time he identified several matters that, in his opinion, needed priority status.

One of these major tasks was to enlarge KHS membership throughout the state, and then increase patronage of the society's learned journal, the _Register._ Most estimates put membership when Chinn arrived at two thousand; that figure climbed almost to ten thousand by the time the colonel resigned his directorship.[1] He wanted to eliminate the "blue-blood" image of KHS. Membership came primarily from Frankfort itself; far western and far eastern Kentucky in the past had routinely been left out in reference to the KHS. One of the staff suggested that KHS directors appoint a person to travel to far-flung Common-wealth communities and actually knock on doors to see if the inhabitants were members of KHS, or if they received the _Regis-_

ter every quarter. The directors considered this proposal reasonable, but were prevented from hiring anyone when the general assembly gave the excuse that seems almost always to be offered: no money.[2]

With little or no help from the state, the KHS recruiters emphasized the "bargain" of being a member: $5 a year or $50 for life membership. This attractive price spiked membership, at least for a while. By October 1970, the society had 9,708 members. A damper was cast on this statistic when someone announced that of that number, 602 were in arrears on their membership fees.[3] One costly error caused severe damage to the KHS budget. Many times when a patron moved to a new residence, he or she listed only the new address for forwarding, without giving the old address. Thus, the individual remained on the books at both addresses. Once the error was discovered, about fifteen hundred names were purged.[4] The creation of a genealogical magazine, *Kentucky Ancestors*, which was supported by Colonel Chinn early in his tenure, brought in additional members, but it was one of the bones of contention between Chinn and Dr. Clark. Clark feared Chinn would turn the whole organization into a DAR-style quest for eminent relatives. He did not.

Among the incentives for joining KHS was delivery of the *Register* each quarter and the chance to obtain a numbered print by renowned Kentucky painter Paul Sawyier. Only members, including newly admitted members, of KHS could purchase these highly prized works of art. They sold for $20 each, considered by most a great bargain. Of course, the problem here was transparent: once someone had his or her prized Paul Sawyier print, there was no need to renew the annual subscription, accounting for numerous losses to the membership. On the other hand, many who did take the opportunity to buy Sawyier prints *did* stay in the organization as a result—and the society grew, even flourished.[5]

The colonel definitely believed that young people should be brought into the membership, a belief shared by Dr. Clark and many other Kentucky historians. Although Chinn, from time to time, chided his juniors for spending too much time on TV and basketball, he loved and respected youngsters. "I have noticed in the past few years," he told a reporter, "so many young people are becoming interested in history." Knowledge of history was especially important for the young; more so than their elders, they were not sure where they were going in life, and before one knows where one's going, according to the colonel, one has to know where one's been.[6] Adults, on far too many occasions, approached ideas and controversies with closed minds. The pursuit of history, both in the classroom and in research activities, needed young minds to establish fresh approaches, to show how the past, present, and future were intertwined. John B. Breckinridge, previous president of KHS, started the Young Historians (YH) program in 1961.[7] Chinn strongly supported the Kentucky Junior Historical Society (KJHS), and worked steadily to increase it membership. By 1970 the KJHS claimed some six thousand members across the width and breadth of the state.[8] Each year the junior historians hold a convention in the Frankfort area: they attend historical seminars, hear speeches from prominent professors in the field, read selected history volumes, and engage in their own historical writing. At numerous colleges and universities in Kentucky, the junior historians helped to inspire and create "history contests," usually in the spring of each academic year. One of the largest of these gatherings is at Western Kentucky University, in Bowling Green.

Colonel Chinn exuded a rosy view of the future. Our grandparents had once been young, prey to all the foibles of youth, and our country did not fall apart as a result. So with today's juveniles: "There's really not that much difference between the ages," he said to a reporter. A twelve-year-old of 1861 was much like a twelve-year-old in 1961, in reference to human nature.

He admitted, however, that he did hold "one thing against the younger generation: 'I envy them.'"[9]

Early in his directorship (1962–1963), Chinn and the Kentucky Historical Society became involved in defending the Commonwealth against the claims of other states about some historical stories and accounts told and retold throughout U.S. history. One concerned the famed long rifle, whose origins were claimed by both Kentucky and Pennsylvania. Chinn, as a weapons expert, explained that the rifle's creators "put rifling" in the barrel of a gun, making the bullet "fantastically accurate."[10] A newspaperman said the long rifle "could hit a man in the head at 70 to 80 yards. You could shoot him in the eye at 50 yards. They [the long rifles] were that good."[11] Pennsylvanians, whom Chinn called claim jumpers, stirred up the matter of the long rifle, asserting that it was first manufactured in the Keystone rather than the Bluegrass State.[12] Chinn knew this was technically true; it was first created on the North American continent by German immigrants in Bucks, York County, Pennsylvania. But he had no intention of acceding to the claim that this "incredible" weapon should henceforth be called the Pennsylvania long rifle. In fact, he did his part to stir up the controversy: "They made Monongahela rye whiskey up there [in Pennsylvania] that was only fit for lamp oil, and it took Kentuckians to show them how to make bourbon. They made a rifle, too, but it took Kentuckians to show them how to use it. They [Pennsylvanians] are the 'claimingest' people."[13] Such an incendiary statement, by the director of the Kentucky Historical Society, no less, most assuredly caught the attention of every professional and amateur historian in Pennsylvania, to say nothing of the governor, William Scranton. His role in the ensuing farcical Kentucky-Pennsylvania "war" over the long rifle was viewed by some as Scranton's preparation for a run for the presidency in 1964. As is well known, he did run in the primaries, but lost out to Barry Goldwater for the Republican ticket.

It was clear to Chinn and his followers that "every pioneer" who came to the Commonwealth's Great Settlement area possessed "a Kentucky rifle" (with nicknames for it like Old Sure Fire, Deer Killer, and Indian Lament); otherwise, there would have been no settlement.[14] "It was an Act of God," Chinn argued, "that the Kentucky rifle was in existence or the pioneer wouldn't have lasted thirty minutes. Guns are the pens history is written with, good or bad." He objected to public attitudes of disdain toward the creation and improvement of automatic weapons. He did not favor gun control on any governmental level: local, state, regional, or federal. Nonetheless, he despised the "gun-nut" who feels that his civil rights are violated if he "can't own seven or eight machine guns and a couple of cases of dynamite."[15]

The major use of this shooting weapon in war was at the battle of New Orleans, in which General Andrew Jackson and his Kentucky Militia effectively used the long rifle against the British. "This country," Chinn practically orated, "has twice had a secret weapon, [and] had they applied it they could have conquered the rest of the world."[16] The first was the Kentucky long rifle and the second, of course, was the atomic bomb.

The discussion over the Kentucky long rifle became so inflamed that Governor Scranton challenged Kentucky governor Bert Combs to a contest to see whether Pennsylvanians or Kentuckians were better shots with the long rife. For historical verisimilitude, he sent his challenge via a delegation of seven horsemen; it took the Pennsylvanians over a month to travel the nine hundred miles to the Cumberland Gap, where they entered the Commonwealth. The Keystone horsemen were met at the border by the Kentucky Highway Patrol, who, after a few minutes of more or less friendly greetings and questionings, let them travel on.

But another person was there to greet the Pennsylvanians at the Cumberland Gap: none other than the director of the Kentucky Historical Society, George Morgan Chinn, himself. He told the horsemen that "dangers" of an "ambush" were "scarce," at least as long as they continuously waved the Kentucky flag he generously supplied them with and "shouted hurrahs for Robert E. Lee."[17] Most ignored the advice; some did not, however—especially when they rode by clusters of citizens standing on the side of the road, casting what the Pennsylvanians considered to be unfriendly glances toward them.

The horsemen were greeted cordially in Frankfort by Governor Bert Combs, who saw to their quarters and food. He was serious about the "shoot-out" challenge: there were tourist dollars involved. Would Kentucky retain them, or lose them to Pennsylvania? The first shoot-out was to be in Reading, Pennsylvania, on September 28, 1963; the second was scheduled for Barbourville, Kentucky, on October 12. Each team of sharpshooters had ten members.[18] Many shoot-out participants hoped that one of the two states would win decisively. If there a tie, there was no telling how long these spectacles would last—at taxpayers' expenses.

There was a win for Kentucky at Reading; Chinn said the Kentucky victory over Pennsylvania was "hands down." Rex Massy of West Liberty, Kentucky, won 150 points out of a possible 155; Robert Rambo from Parkersville, Pennsylvania, achieved 124 points out of 150. Their targets were swinging jugs and turkey heads. Kentucky finished with 975 overall points, Pennsylvania trailing with 839.[19] The Barbourville shoot-out ended with another Kentucky victory. Ever since then, small meetings between sharpshooters of the long rifle from Kentucky and Pennsylvania have taken place each year. But, thanks in large part to George M. Chinn and Bert Combs, the title of this extraordinary weapon remains the *Kentucky* long rifle—and is apt to stay that way.

After putting aside the Kentucky long rifle—at least temporarily—Colonel Chinn turned his attention to another matter of

considerable importance, a subject that occupied him throughout the whole of his tenure as director of KHS: Kentucky icon
Daniel Boone, about whom Chinn had mixed feelings. One day,
perhaps while driving to work in his multicolored car, Colonel
Chinn conceived an idea he deemed brilliant. When he floated
the plan before his colleagues at KHS, he was gratified that all
of them fully supported it. Each year since the late nineteenth
century, a Daniel Boone Day had been celebrated, with activities centered primarily in Frankfort. Chinn wanted to extend
the festivities to other localities—including, of course, Harrodsburg. Naturally, the Kentucky Historical Society always participated in these festivities. Volunteers went door to door out into
the communities several weeks before the event to solicit funds
for the celebrations, while others gave their time to building
stages, sheds, and even replicas of frontier houses to show what
life was like during the Boone era. Colonel George M. Chinn,
who had always publicly respected Boone but privately argued
that he was "overrated," thought it would be great if the Young
Historians took the lead in preparing for Boone Day. Most of
the youngsters (and grownups as well) in the Big Settlement area
approved the idea.

On July 6, 1962, KHS board members were told that
$160.35 had been collected for the next Boone Day, which was
almost always scheduled for late June of each year.[20] In April
1964, Boone Day funds came to $378.90, enough to pay for the
festivities and enjoy a balance of $149.05.[21] KHS board mem-
bers suggested that the Young Historians be given permanent
charge of Boone Day. Allen Trout of the *Louisville Courier-
Journal* volunteered to edit all the essays YH members wrote
for the occasion.[22] Clearly, the Young Historians had an impact
on both fund-raising and the historical content of the yearly
Boone gala. The occasion provided a wonderful opportunity for
young people to learn about one of Kentucky's most romantic
and famous figures.

Boone Day on June 25, 1966, was a great success, helping to increase enrollment in the KHS itself. The late 1960s and early 1970s, however, were not good years for the Boone Festival in Harrodsburg. Scandals in Washington, D.C., such as Watergate, took attention away from academic pursuits, and a presidential resignation (the first in our history) dampened the country's morale. The Arab gasoline embargo added misery to the country as a whole. Therefore, the Young Historians program was put on hold, and Boone Day was reduced to a fragment of what it had been.

Fortunately, in the years to come, both the KHS and the YH were restored almost to their original strength and vitality. In the late 1970s, for example, the Kentucky Young Historians joined with similar clubs in Indiana and North Carolina to pursue the study of history. By 1987 membership had grown to 3,700, with 175 different clubs throughout the state. KJHS also celebrated its twenty-fifth founding anniversary, which gratified and justified many old-timers, including Colonel Chinn, who had had the foresight to develop and support this organization in the first place.[23]

Many people felt that Boone needed not only to be memorialized but treated as semi-sacred as well, and spent their time (and sometimes money) keeping Boone before the public eye. They succeeded, but many times in ways they did not desire. The board, and members who followed its activities, wanted the Federal Communications Commission (FCC) to do something about the scandalous sensationalizing of the Daniel Boone story on television. Joe Creason of the *Courier-Journal* volunteered his time and writing talents to try to convince the FCC that the Daniel Boone portrayed on TV was "too highly dramatized."[24] The FCC, this group felt, should require the show's producers to make disclaimers both before and after the program informing the public that these episodes were not based on historical or biographical facts. The board members did not want to repress the show; they simply desired to notify viewers of *Daniel Boone*, particularly children, that the show was fictional, not the truth about one of Kentucky's most significant icons. Creason's efforts, however, were "completely wasted" with regard to the "real" Daniel Boone versus Fess Parker (the actor who played him).[25] The networks wanted *entertainment!* If that entertainment should coincide with the happy truth of authenticity, all well and good; if it did not—well, then, truthfulness had to fall by the wayside.

In addition to the troubles of falling funds and the condescending attitudes of TV executives, KHS and YH began to have problems finding keynote speakers for Boone Day. Those to whom offers were made included Jesse Stuart, John Jacob Niles, Jean Ritchie, Dr. Thomas Clark, and Daniel Boone's descendant Pat Boone, the singer.[26] For two or three years, Boone Day depended on local talent, which, to the surprise of many people in the Great Settlement area, in many instances turned out to be as good as or better than "professional" presenters.

In public speeches and dialogues at the KHS in Frankfort, Chinn lauded the life and times of Daniel Boone, and he went on record as one of the strongest supporters of Boone Day. Privately, however, he argued that Boone had a wonderful "agent" in the person of John Filson, after whom the Filson Club in Louisville is named. "If it hadn't been for Filson, who wrote a book about early Kentucky, you'd probably never even heard of Boone," he told interviewer Ellis. He explained that on June 16, 1774, James Harrod, originally from Pennsylvania, led a group of settlers into what became Harrodsburg. On May 25, 1775, Chinn continued his argument, "Boone's party reorganized and arrived at a settlement that came to be called Boonesborough, acclaimed by one Boonesborough Chamber of Commerce after another through the years to be the 'oldest' town in the Commonwealth of Kentucky." Not so, said Colonel Chinn; Harrodsburg held that distinction.[27]

Boone and Boonesborough, the KHS director insisted, *would* have been first if not for some unfortunate and tragic incidents that occurred in Tennessee along the Clinch River. Boone's party, heading for Kentucky, came upon a large group of Shawnees having their breakfast. Boone ambushed the Shawnees, and one of his sons was killed. (Of course, other members of Boone's party suffered losses in this attack, but newspapers and books, both romanticized fiction and otherwise, concentrated on Boone.) The wives of the Boone party refused to move any further until the territory was at least halfway pacified. This delay in movement explained why Harrodsburg lays claim to being the oldest governmentally authorized settlement in Kentucky, first from Virginia (of which Kentucky was a part) and then from the Commonwealth of Kentucky. Harrodsburg was established, Chinn adamantly claimed, on October 10, 1785, settled essentially by veterans of the French and Indian Wars. Boonesborough, Chinn said, was never a settlement because it did not meet certain requirements imposed by the state legislature, such as having fresh springs every few miles for inhabitants, both for those already settled and those on the way. It was legally a town but not any kind of authoritative place, such as a county seat; it was that way then, and that way now.[28] Chinn and Cotton's cave out on Highway 68, Chinn joked, was the only cavern in the entire Commonwealth where Boone had *not* spent a winter.[29] Of course, Boone had never claimed *any* of the places where he and his party lingered for a time on their way to what became known as Boonesborough. Chinn was simply being facetious.

Walking down a hallway one day at the KHS building, Chinn noticed a new display hanging on a wall. He recognized it immediately as a Boone rifle. He couldn't resist; he took it and left work for the day. The KHS staff searched frantically for the missing rifle for a few days, only to discover that it had been mysteriously reinstated to its place. Chinn, asked if he were somehow implicated in the "case of the missing rifle," sheepishly admitted that he had "borrowed" it for a while. He had taken the old firearm to Shepherdsville and fired it. He fired it left-handedly, although the gun had been manufactured to be used by right-handers. Chinn was right-handed for most purposes, but he fired a gun and swung at a baseball left-handed. The weapon's kickback left powder burns on Chinn's face; they were not deemed serious at the time but required cosmetic surgery in the years ahead, reminding Chinn to be extra careful with old firearms.[30] Besides, he had already experienced enough gun powder and its subsequent noises in the military to know what kickbacks can do to face and ears. The board decided that "legitimate" members were authorized to take items away from the museum, *if* they'd let staff workers know they had taken them and indicated how long they meant to keep them.[31] Ultimately, this practice was discontinued; *nothing* was to be taken from the Kentucky Historical Society by staff, high or low.

An alternate story of the missing Boone rifle has it that Chinn wrested it from KHS and took it to Elizabethtown in Hardin County, supposedly for the dedication of a highway. "At the sound of the shot, the governor [John Y. Brown Jr.] was supposed to cut the blue ribbon." The rifle Chinn held had not been fired for two hundred years; nevertheless, "I pulled the trigger and 'Wham.'" Chinn had forgotten, he shamefacedly said, that "the longer you fool around with guns, the more careless you become."[32] He was embarrassed by the incident of the "borrowed" Boone rifle. Did he fire one shot at Shepherdsville? And/or one at Elizabethtown? Did he suffer powder burns at just one city or at both places? Whatever, his escapades with the Boone rifle contributed to his facial and bodily ailments.

Chinn was incensed by rumors that authorities in Missouri had not shipped the real bones of Daniel Boone back to Frankfort for reinterment but "instead had substituted those of an unknown person." Chinn's characteristic response was "Bunk." One of Boone's descendants and eleven other residents and relatives of Daniel Boone were present to affirm what they considered to be a true disinterment in Missouri and a reburial in Kentucky.[33] The argument over whether they are truly Boone's bones, however, goes on to this day in Kentucky.

Born in Pennsylvania (which some Keystoners never let the Bluegrass forget), Boone was a peripatetic man. He'd be in the Tennessee area of the Clinch River, then he'd show up in the tiny settlement of Boonesborough, Kentucky, and then he'd surface in Missouri. His travels never stopped until Boone's body was supposedly moved back to Kentucky. And even then, many Kentuckians thought the Missourians were trying to pull a fast one on them.

In due time, it became clear that it wasn't Missouri that posed the problem about Boone's burial site in Frankfort. Boonesborough's city officials thought he should be reinterred in one of that city's cemeteries. Thus began a conflict between Frankfort and Boonesborough that, if it never quite reached the level of a "feud," for which Kentucky is either famous or infamous, it came close. Boone had spent more time in Boonesborough than in Harrodsburg, and certainly more than in Frankfort. It seemed reasonable to the city council of Boonesborough that Boone's final resting place be the city that bore his name.[34]

KHS director George Chinn strongly opposed moving the body to Boonesborough. To move the grave, he said, would take a legislative act "and then a very good-sized army to enforce the act." Referring to Frankfort's cemetery, high on a hilltop overlooking the capital, Chinn noted that "Boone's up there in good company." The Frankfort Cemetery, he said, was a "Kentucky Arlington" or "like England's Westminster Abbey." Furthermore, why should Boone's bones be disturbed once again? Why not transport Abraham Lincoln from Springfield to Hodgenville? Or Confederate president Jefferson Davis from Richmond, Virginia, to Fairview, Kentucky? The number of visitors to Boone's tomb in Frankfort and the historical arguments given by Chinn and his colleagues at KHS helped considerably to keep Boone's grave undisturbed.[35]

Boone had resided at many places in Kentucky and had shown up for short visits at numerous campsites, causing some historians, both professional and amateur, to wonder about the total mileage Boone and his followers had walked in their Kentucky odyssey. And this is where the Historical Markers program, sponsored by KHS, came into play.

The Kentucky General Assembly and the Kentucky Historical Society both wanted in-state and out-of-state motorists to know the historical significance of the areas through which they were driving. This reanimated the Historical Markers program in the early 1960s under the guidance of Director Chinn. A previous such program, before Colonel Chinn's involvement, dated back to 1947, when Governor Earle Clements and one of his aides, Vear Mann, acted positively on a suggestion by KHS treasurer Bayliss Hardin.[36] Its effectiveness was variable throughout the years. That Chinn was now in charge of this enterprise caused Dr. Clark to grumble: he thought the "free-wheeling" Chinn ignored the old committee, or at least Dr. Clark's part of it.[37] Apparently, Chinn had advertised in newspapers far and wide some "extravagant" promises of markers to communities—as long as they had the "right" (meaning, in this instance, "correct") politics. "If citizens are to serve the state," Dr. Clark exclaimed, "you have to quit acting like a bunch of partisan politicians on these matters."[38]

At a board meeting, Chinn announced that 82 percent of all historical markers in the state were placed in Central Kentucky, "chiefly in the vicinity of Lexington." This was the very kind of statistic Chinn deplored. As he had with membership drives, he wanted to include Eastern and Western Kentucky and the Jackson Purchase, making sure that the residents in those areas were fully represented by the Markers Committee. This lack of equal representation caused citizens in these areas to view the

KHS as "elitist," made up primarily of "historical amateurs" or "bluebloods" from Lexington to Frankfort, Harrodsburg, and other places in Central Kentucky. Chinn argued that for the KHA to survive and flourish, this image of gross inequality had to be corrected.[39]

At another board meeting, Dr. Lowell Harrison, professor of history at Western Kentucky University, moved that "proof of accuracy" be required for all markers installed in the state. Chinn suggested a committee composed of professional *and* talented amateur historians from all over the Commonwealth to read inscriptions and verify their accuracy.[40] The committees were appointed, but their personnel came essentially from the local populace rather than from statewide historians. The present director of the Historical Marker program, who works for the KHS in cooperation with the State Transportation Department, is Becky Riddle. In the early summer of 2013, she oversaw the installation of the twenty-four-hundredth historical marker.[41]

A marker application will not be approved if it will create a traffic hazard, in the opinion of the Transportation Department, if it is not acceptable to the property owner where the marker is planned, or if the local government has objections. A marker will not be approved for a living person, for one who died fewer than twenty-five years ago, or if its subject is purely genealogical or family related. A cemetery will not be accepted unless it has at least one historical figure interred there; a church is off limits unless it has had a congregation going back at least 150 years, with constant attendance in a building that is at least a century old, and has an architecture that is unusual and unique for Kentucky. As of 2013, a marker with the same text on both sides costs $2,300; one with different text on each side comes to $2,500. All markers are funded locally, and the wording on each side has a maximum of ten lines with thirty-three letters or numbers, including spaces.[42]

Historians at KHS carefully edit all markers. If an error is found, the process to correct it becomes somewhat clouded. Almost always, it is a case of informing the local Marker Committee, which sees to it that the necessary corrections are made. In many instances, however, mistakes remain on the board.[43] On the whole, however, the Historical Marker program has been a spectacular success.

Colonel Chinn would have rejoiced in the success of the program that he was so instrumental in starting some fifty years ago. History-minded travelers throughout the Bluegrass State find their trips greatly enhanced by greater knowledge of important events and people in the places they visit. Kentucky travel writer Gary West summed up the significance of the historical markers: "I've noticed that the older one gets the more time they take to read the roadside markers." Furthermore, West avowed, historical markers "validate their existence as a historical roadmap to those who take time to read them."[44] The Historical Markers program ultimately has been one of the KHS's most successful. It has led to innumerable visits to historic sites in the Commonwealth, to say nothing of a brisk expenditure of tourist dollars.

Not the least of Colonel Chinn's ambitious programs dealt with advancing a scholarly press and funding an elaborate governors' papers project (and in doing so, supporting a massive microfilm operation). Numerous authors came before the society seeking publication of manuscripts about Kentucky. For example, G. Glenn Clift (one of the officials at KHS) presented *The Private War of Lizzie Hardin* for the publication committee's consideration. The book, edited by Clift, is the diary of a young girl kept during the Civil War. The Kentucky Historical Society gladly published the work.[45] James W. Sames III's book *Four Steps West*, published by KHS, was popular during the bicentennial celebrations in the 1970s.[46] Willard Rouse Jillson, another society officer, announced to the publication committee that he was writing an "all inclusive" history of Frankfort, and suggested that the society pay for its publication, a proposition that caused a few grumbles throughout the historical community.[47]

Closely connected with the publishing arm of KHS was the Kentucky governors' papers project. This involved compiling and cataloguing documents, and assisting the University Press of Kentucky in its plans to publish the papers. In this respect, microfilming of these documents was crucial to the success of the project. By October 1963, Joan Brooke-Smith, the director of microfilm, and her staff were "ahead of schedule, having in fact, reached the half-million mark in duplicating gubernatorial records."[48]

Miss Brooke-Smith was something of an institution at KHS. Originally from England (her father, a British sea captain, brought his family to the United States to escape the London blitz in 1941), she ended up in Kentucky because she loved horses and racing. Brooke-Smith was with the historical society for twenty-two years, and Chinn praised her not only for her work in the microfilm department but also for her role in turning the society into a friendly, welcoming place for researchers and genealogists. She also worked at the nearby Shaker Village of Pleasant Hill, confirming land grants, among other things, for this religious order. Chinn reported that when a college student showed up for a "dig" at Pleasant Hill to see if a certain building he'd read about had actually existed, Miss Brooke-Smith's excellently kept files provided him with "the building, the number of feet, the number of jousts, and the type of nail." Brooke-Smith's work was valuable in preserving historical accuracy, Chinn asserted; her research proved false the rumor that Shakers would not own mules because they were hybrid animals: not only did the Shakers own mules, "they had two huge mule barns." Brooke-Smith, Chinn said, made a career of "getting this Shaker stuff together," and she contributed immeasurably to the success and fame of the Kentucky Historical Society.[49]

Miss Brooke-Smith—along with Colonel Chinn and other

staff workers—feared that affiliation with the University Press of Kentucky (UPK) in dealing not only with the governors' papers but other manuscripts as well might cause the Kentucky Historical Society to lose some, or perhaps all, of its autonomy.[50] Apparently, KHS had considered other presses, some out of state, for publishing the governors' papers. The KHS officials were told—much to their dismay—that there was "only one printer in the state authorized to 'take care of the typography' and indexing of manuscripts published either in whole or in part with funding from the Commonwealth of Kentucky."[51]

KHS officials wanted to publish a total of eighteen governors' papers (Combs, Chandler, Nunn, and Breathitt were among the state's governors to grant quick approvals), with one volume coming out every two years, for a total printing and publishing cost of $26,000. If KHS were to be essentially forced to use UPK for the governors' papers, the society wondered what the implications would be for other current or future publishing projects. Authorities at the University of Kentucky assuaged such feelings by saying that UPK-KHS affiliation in no way meant that every KHS publication from now on *had* to come through the UPK. The university promised that it would advertise the governors' papers and sell copies to all the university and college libraries in the United States and to "major universities" throughout the world.[52] This would, the university believed, more than compensate KHS for its investment in the governors' project.

Still, questions remained. What about editorial guidance? Who had the final say-so in reference to actual publication? The Kentucky Historical Society? The University Press of Kentucky? Many in the society thought they should have "final rights" on manuscripts. If the KHS board came up with a controversial decision about a manuscript, it should be turned over to an editorial board (one more hassle for the hapless author?) at UPK, which would have final authority on the matter.[53] In the midst of these discussions, someone in the State Finance Department transferred $25,000 that had tentatively been earmarked for KHS to UPK—without the permission or knowledge of Colonel Chinn or, for that matter, appropriate spokesmen at UPK. This outlay was to cover the costs of publishing the governors' papers. Chinn told the board that it was "not his intention" to create dissension with anyone (at least on this occasion), but he was "responsible" for the funds of the historical society, and he needed guidance on how to handle this situation. Ultimately, he and the board let the matter rest. Under the circumstances, rumors abounded; one of them had the publication department at KHS being moved to the University of Kentucky itself, an alarming prospect for KHS personnel.[54] The history society is still in Frankfort.

The clincher in this whole deal came with UPK's request that KHS pay a $1,000 entry fee for an affiliation with UPK. At first Chinn believed that this "last-minute" provision was nothing less than extortion. When he found that this was only a one-time fee, which all other members of UPK also had to pay, he relented, though there were still some resentments in the publishing department of the Kentucky Historical Society.[55] Since that time (circa 1969), KHS has published a few books on its own (as has the Filson), distributed by UPK, and recently KHS and UPK jointly published a book of scholarly interest: James Ramage and Andrea Watkins's *Kentucky Rising: Democracy, Slavery, and Culture from the Early Republic to the Civil War* (2011). "This recent joint publication," says one KHS official, "makes the publishing relationship between KHS and UPK more visible and transparent to the public."[56] Once or twice a year representatives from all the educational and historical institutions in the consortium for which UPK is the "scholarly publisher for the Commonwealth" send a delegate to Lexington to discuss and vote—yea or nay—on selected manuscripts. On the whole, over the years, the consortium has worked well, with professionalism and courtesy from all its members.

In addition to all the major projects Chinn and staff worked on throughout the 1960s, there were numerous little or mundane things to be seen to. KHS was expected to participate in most if not all historical celebrations within a wide swath of the Bluegrass area. At one board meeting, a major discussion occurred over raising a flower fund, which seemed to interest Chinn as much as Daniel Boone had.[57] A letter arrived from the government of South Korea, wanting KHS to find a relative of a U.S. casualty in the "police action."[58] KHS found the relative. In December 1964, Mary Hinton of Versailles, Kentucky, offered her house as a gift to KHS,[59] and famed author Harry Caudill wanted to contribute a cabin in Letcher County to the society. These offers and several others were gratefully accepted by KHS.[60]

The board fully discussed "no smoking" policies at the library; gradually, they went into effect. Chinn himself had no problem with designating certain places as "nonsmoking."[61] While he was in the Marine Corps, he was fond of cigars, which he chomped and chewed on while working in ordnance departments. The society was called upon to clean up, beautify, and rededicate the Jefferson Davis Memorial at Fairview Cemetery in Western Kentucky.[62] And the talk seemed never-ending about building museums, historical and military, in Frankfort and other selected cities in Kentucky. Some thirty-four states were interested in preservation societies, along with six foreign countries. All of these sent researchers to Frankfort; Chinn, Brooke-Smith, and the research staff did all they could to furnish these visitors with the large quantities of the microfilm and other research materials they wanted.[63] Many people also wanted to talk with the KHS director—time-consuming affairs—for they had heard he was an expert in military weaponry. The paperwork involved in these projects was voluminous.

No wonder, then, that in late 1969, Chinn told his friends that his physician had ordered at least a three-week vacation, in which he was to make absolutely no exertions.[64] He didn't get around to taking this advice for at least another three years. Earlier in the year, January 15, he had turned sixty-seven, yet, his schedule would have challenged a much younger person than he.

Perhaps his fatigue and impatience contributed to an unhappy incident. Chinn turned away an African American who wanted to donate a significant collection of nineteenth-century black newspapers to KHS. (African American collections, including newspapers, were scarce among the repositories of Kentucky history.) A historian at KHS "was horrified" to hear Colonel Chinn talk about this incident. "They were probably stolen," Chinn explained of his refusal to accept the newspapers—too nonchalantly, most of his audience believed. One of his colleagues noted that Chinn may have been "devoted to KHS, interested in history and an advocate for some new programs," but he did not have the "academic's mindset regarding collections."[65] This action toward the African American was totally out of character for Colonel Chinn. At no point in his long military service had he been accused of racism.

In fact, he strongly supported school integration during the 1960s, and helped state police commander James E. Bassett ("Ted") control racial situations that threatened the peace of the Commonwealth's schools, colleges, and universities. The police Special Forces (they had no SWAT teams at the time) were short on weaponry. Chinn heard about this problem and immediately contacted an official—one of his friends—at the Mechanicsburg, Pennsylvania, navy arsenal; in fact, it was the "collection point" for naval weapons left over from World War II. The firearms were thirty-caliber carbines, which pleased Commander Bassett, for they were useful for crowd control and for breaking up riots. "We got them, some fifty or sixty of them," exulted the commander, but "we never put them to use. The deterrent was enough to discourage citizens from riotous behavior over the desegregations of the Kentucky school systems." The police Public Relations Department in numerous press releases made sure that Kentucky newspaper readers knew about the additional weaponry in the hands of law enforcement officials.[66] Many would-be rioters stayed at home when they learned of this situation. Thus, Kentucky's integration was more peaceful than that of most other states in the American South. This happy result was thanks in part to Colonel George M. Chinn's willingness to help the state police achieve racial equality peacefully. Commander Bassett and Colonel Chinn remained good friends for the rest of Chinn's life.

Chinn was tired, sometimes to the point of stressful exhaustion. He needed to take the advice of a three-week vacation, to rest, do some reading, maybe even a bit of writing, perhaps think about his less than gracious and abrupt refusal of the African American's donation. At any rate, he decided in 1973 to retire as the director of the Kentucky Historical Society. After his thirteen years as director, he would now spend a few more years in the less demanding position of deputy director. One big lure for Chinn to retire (he was now in his seventy-first year) was the opportunity to direct the military museum being set up by KHS and the State Department of Military Affairs, scheduled to open in mid-1974. "There isn't a modern weapon up there that I haven't had some contact with," Chinn said. With his knowledge of weaponry, he was certainly qualified to direct the activities at a military museum. Chinn took a drop in pay as deputy director of KHS; his annual salary went from $13,200 to $10,872.[67] Chinn happily accepted the changes, for it would relieve him of many of his duties at KHS and allow him to look forward to building and nurturing the military museum.

Chinn's successor at KHS was William R. Buster, a fifty-seven-year-old retired brigadier general with whom Chinn had always gotten along. In fact, while Chinn was deputy director under Buster, many observers spoke of a "joint directorship";

their actual titles were fairly well ignored by both men. The two worked in tandem for the good of the Kentucky Historical Society. They both shared in a part of the history of the Commonwealth: Chinn's grandfather and Buster's great-uncle had both been present at Goebel's murder.

Of course there were numerous reactions to Colonel Chinn's retirement, not only at KHS but at UPK and the academic community throughout the Commonwealth. Naturally enough, everyone wanted to know what Dr. Thomas D. Clark, at this time a professor at Eastern Kentucky University, thought about it. He gave this enigmatic statement: "There was so much that could have been done if they had used the trained talent in this state."[68] Unfortunately, Dr. Clark did not explain who he meant by "they," what it was that "could have been done," or the nature of the "trained talent." What could have been done beyond Colonel Chinn's efforts? Clark himself, in the past, had lauded Chinn's endeavors to increase the KHS membership as well as make the society a statewide organization rather than one simply affiliated with Frankfort. One wishes that he had defined his phrase "trained talent" in particular. Did he mean professional historians in large university history departments? Or talented amateurs without any connections to higher education?

He did give an opinion of Chinn's departure, although it was somewhat veiled. Speaking to interviewer William Marshall, Clark, in praising Buster, could not help but make an unstated contrast with Chinn. "Bill Buster was not a historian in the true sense of the word." That, of course, had pretty much been his chief complaint about Colonel Chinn from the very beginning. Clark went on regarding Buster: "But he was a good manager. He was a good man with personnel and he had vision."[69] Such assessments could as well have been made about Chinn.

Would Chinn have accepted, even savored, these encomia from Dr. Clark back in 1959 when Chinn took charge of the KHS? Probably, but he had enough faith in himself to rise above

critics and get on with the job. After a few years as deputy director of KHS and leader of the military museum, he decided that
he might just retire from *everything*, go back to Harrodsburg,
and live with Cotton in their magnificent house on top of the
hill. In the event, though, he seemed to become involved in
more activities in the late 1970s and on into the 1980s than ever
before. He had long been a celebrated hallmark of the Big Settlement. It stayed that way for the rest of his life.

8

∾

The World According
to Chinn

If Colonel Chinn thought he could just walk away in comfort, peace, and serenity from KHS and return to Mercer County to live a trouble-free life with Cotton in their palatial home atop a hill overlooking the palisades of the Kentucky River, he was sadly mistaken. One of his earliest projects back in Harrodsburg dealt with activities of the Mercer County Humane Society. He always had a soft touch for animals, especially dogs. Prior to his and Joan Brooke-Smith's involvement, the policy of the humane society was to capture, kill, and then dump unwanted cats and dogs into the local quarry. Not long after Chinn became active in the humane society, there were dramatic turnarounds. For one thing, he talked the Mercer County Fiscal Court into a sizable grant each year to hire additional staff and maintain the newly constructed humane society building.[1] This building became one of the first "no-kill" establishments in the state.

Colonel Chinn believed the Harrodsburg/Mercer County Fiscal Court, in cooperation with the Bicentennial Committee, should sponsor a fund-raiser to benefit the humane society as well as activities such as Boone Day and Fort Harrod. Chinn concluded that the best project would be a collection of recipes submitted by local residents. He felt this was sure to sell. And it did, especially when word got out that the colonel himself had contributed two "mouth-watering" recipes. Residents of Harrodsburg had not known that Colonel Chinn was also a chef; after they tried his recipes, they still didn't know.

His first recipe was for beer cheese; the major ingredients were five pounds of Longhorn cheese, one-half a head of garlic (three "toes"), one jar of yellow peppers, two bottles of flat beer (the colonel advised pouring the beer into a receptacle to set a spell), half a teaspoon of salt, and an ounce of Frank's "Red Hot Sauce"—or two, "if you're brave." After blending the cheese, garlic, peppers, beer, salt, and hot sauce, the mixture "will be a dark brown gloop." (That statement alone should have warned people off trying the recipe.) A large helping of horseradish, the colonel advised, should be added at the last minute. The mixture should be eaten with crackers and a "soothing, nonflammable drink."[2] According to a goodly part of the local community, this recipe from the beloved colonel tasted as awful as it looked.

The second recipe the colonel submitted, however, received wide attention, even acclaim. Listed under the main dishes category, it was called Colonel Chinn's Famous Bully-Beef Frappe, and it was nothing short of sensational. It called for one pound of cooked tough beef, cut from around the hock. The cook was to put this into an earthenware container and then pour over it a quart of one-hundred-proof bourbon and one-fifth of a bottle of rum. The only way to check the dish's progress was to take sips from the mixture until the preparer deemed the concoction palatable. Then the dish should be placed in the refrigerator for at least twenty-four hours. During that time there should be "generous tastings" to ensure the preservation of flavor. "At least two hours before company comes, remove from refrigerator so that serving can be at room temperature. Having decided it is ready, remove the bully beef from the broth and place on a wooden slicing board. Divide into equal portions and garnish with watercress. Call in all your pets and feed the beef to them—it was never meant to be eaten anyway. However, don't let anything happen to that broth because it'll cure anything that ails you!!"[3] Readers loved this recipe, although it is unknown if anyone ever actually tried to make it. But the colonel achieved his mission: the recipe book was a runaway best seller in Harrodsburg and Mercer County.

The recipes may have inspired Colonel Chinn, now KHS deputy director, to create in 1980 the first of what was intended to become an annual event, a "talk meet" to bolster morale at the historical society. Sessions were scheduled for "discussing, discursing [discoursing?], and dis-emboweling" all the problems (be they human or otherwise) in this universe and all others in the galaxy. The first talk meet's judges remained anonymous for their own safety. This talk meet's motto was "Semper Fidelis"; it essentially revolved around Marine Corps history, and the intention was "to out-talk the women of this world." The talk meet's first champion was none other than "Col. George Chinn, Esq." Regrettably, the advertising broadsheet does not mention Chinn's opponents on this occasion, or who won the "confrontation." The sponsors were quick to point out that "this certificate was done by an independent contractor" to avoid that taxpayer money was being spent on "frivolous" historical exercises.[4] No evidence exists that other sessions of talk meet ever took place, or if the first one actually improved morale. On the whole, morale at KHS was healthy; most personnel said it was a good place to work.

Some KHS personnel, however, were not as happy with Chinn as those who were reading and relishing his recipes and attending talk meets. For example, one of KHS's books, *Kentucky: Settlement and Statehood*, was poorly indexed. The indexer did not employ any kind of cross-indexing, so that under "Boone, Daniel," for instance, there was a listing of some two hundred separate page numbers. Another indexer had to be employed to go back to fix the problem, a time-consuming and tiresome process.[5] The lack of a cross-index was, of course, no crime, but it was a sign of nonprofessional treatment.

More serious were issues of the book's authorship and copyright. Libby Frass helped with the research and Dr. Hamilton Tapp wrote some of the post-1792 chapters. In later years, Tapp "mildly suggested" that he should have "received more recognition for that" than he did. Furthermore, Colonel Chinn published another book after he left Frankfort, *The History of Harrodsburg and the Great Settlement Area* (1985). "But most of the early part of that book reprinted, verbatim, portions of *Settlement and Statehood*, which was under KHS copyright."[6] Apparently, nothing was ever said to the colonel about this matter.

Colonel Chinn's book problems continued when he set about to publish volume 5 of his machine-gun series, which he thought of as a recap of the first four volumes. (Volumes 1–4 were no longer classified by the military.) Chinn had been in negotiations with UPK to publish the fifth volume. Then, at least according to one version of the story, a research institution at the University of Kentucky known as Spindletop became involved. This organization apparently applied for publication monies for the volume from various foundations and philanthropic institutions, all to no avail. Chinn decided to bankroll the volume himself, and it was brought out by an agency called Ramp, Inc. Reportedly, UPK clamped down on any and all information that had been collected up to the time of Spindletop's final failure to obtain funds, and that is where the matter stands today; there are documents from that era closed to everyone, including both amateur and professional historians. Chinn was never certain whether the directors at UPK even knew that volume 5 ever came out, let alone what was in it.[7]

Comments in some of Chinn's other books raised historical eyebrows throughout the Commonwealth. These seemed to be motivated by Chinn's tendencies to shoot from the hip when he felt he had something important to say. But in effect, this attitude helped to elevate him into the highest echelons of 1980s historical writings about Kentucky. An article in Danville's *Advocate-Messenger* summed up this thought: "Colonel George Chinn may not be the smartest man to graduate from Centre College, but he might be the most successful." Given all of his travels throughout World War II, Korea, and Vietnam, he was clearly a "Roads Scholar," said the Danville newspaper. "He can be a leader in any group."[8]

Despite the accolades for his work, however, some reviewers, scholarly and otherwise, continued to look into what they considered the more egregious statements in Colonel Chinn's books. One concerned the question of where the original Kentucky settlers came from. They were heavily Scotch-Irish, Chinn argued, and they *did not* come through the Cumberland Gap into Kentucky. Some did; most did not.[9] Chinn said that it took another fifty years for the Johnny-come-latelies to start coming through the Cumberland Gap into Kentucky; now, if one were going in the other direction, from Kentucky to Tennessee, for example, the migrants overwhelmingly used the gap from one state to another.[10] The only way one could enter the Great Settlement was by the rivers. Settlers came down the Ohio River from western Pennsylvania (mostly the Pittsburgh area), took a turnoff, and branched out on Salt River. Early comers walked up this branch and got to the "high ground."[11] They could look down into the valley and envision a new town or, ultimately, city. This vista became Harrodsburg. There was a large, all-weather spring in the area, which encouraged settlement.

Many textbooks and various other writings had claimed that Harrodsburg was populated first by veterans of the American Revolution. Chinn discounted this, saying that it was veterans of the French and Indian War (1756–1763) and their descendants who first arrived. Harrodsburg, he always reminded his audiences, was formed in 1774 and the first shots of the American Revolution were not fired until 1775.

He loved to talk about Fort Harrod and give "Red Arrow" tours (each point of interest was painted a bright red). Though he said at one time that there were no Indians at Fort Harrod when the first settlers came, he did speak of "differences" between the Caucasian and Shawnee inhabitants. The roofs were slanted inward, he noted, so that if the Indians set fire to the place, residents could put it out without exposing themselves to arrows and bullets.[12]

Chinn's most controversial statements about early Harrodsburg dealt with renowned military man General George Rogers Clark. In a room called the Blockhouse at Fort Harrod, Chinn claimed, General Clark plotted his northwest expedition during the Revolution. It was "one of the 10 greatest military decisions of all time," the Colonel argued.[13] Chinn adamantly asserted that Clark's northwest expeditionary force, though outnumbered by the British, "saved" much territory for the United States, namely, the future states of Ohio, Michigan, Indiana, Illinois, Wisconsin, and Minnesota.[14] General Clark, the "Hannibal of the Great West," began his campaign against the British with fewer than two hundred men on June 24, 1778. Ultimately, Clark and his soldiers (many of whom were equipped with Kentucky long rifles) captured Kaskaskia, Cahokia, and Vincennes. If these battles had ended the other way around, Chinn stated, Kentucky would probably never have become a part of the fledgling United States. Harrodsburg was the primary base of operations for the defense of the frontier during the Revolution. Had Harrodsburg fallen, all of the northwest frontier would have been retained by the British, Spanish, or French, depending on who won the next war. One effect of this possible setback "might" have "resulted in a long narrow strip between the Appalachians and the Atlantic Ocean" becoming the United States, with some other country, more than likely a European one, ruling the rest of the North American continent.[15] Because of General Clark this, happily, did not happen. Chinn said that in "surprising

British and Indian forces at Vincennes," Clark and his 157 men "accomplished the equivalent to a battalion from Harrodsburg going to Europe during World War II and capturing both Hitler and Mussolini."[16] Clark "forced the British border from the north bank of the Ohio River to the center line of the Great Lakes," thus imposing further territorial losses on the British. Chinn continued with some speculations of what a victory for the British would have meant to Harrodsburg and surrounding areas. Among other things, it would have hampered, if not entirely prevented, the ultimate three-thousand-mile drive to the Pacific, "which historians have labeled this country's 'Manifest' Destiny."[17]

Chinn drove both amateur and professional historians up the wall with his hyperbolic statements. His conclusions were considered unjustified overreach in many quarters. When told that his ideas about first settlers, General George Rogers Clark, and especially Manifest Destiny (because this doctrine is cherished by so many Americans) were "provocative," Chinn replied, "If you can't be controversial—and back it up—why bother to do anything?"[18]

There was no good relationship between Chinn and Dr. Thomas D. Clark to begin with, and Chinn's interpretations of General Clark, the Northwest Territory, and Manifest Destiny only widened the gap. Dr. Clark, the best-known and most popular historian in the state, scoffed at Chinn's remarks, echoing other professional historians' opinions about Chinn's work. With the state's leading amateur historian and the state's most intellectual historian at odds with one another, the great masses who were interested in historical studies—or perhaps would have been—once again began to feel left out in the cold. The loser was the valid status of the study of Kentucky history.

Less controversial than his statements on General Clark and Manifest Destiny but still viewed with disfavor were Chinn's remarks on Kentucky in the Revolution, the first and last battles of that conflict, the Bluegrass State's lack of participation in creating the Declaration of Independence, and its involvement with the writing of the Constitution. On October 10, 1774, Chinn claimed, Point Pleasant, Virginia (now a part of West Virginia), was the scene of a confrontation between Loyalists and Indians at the mouth of the Kanawha River, in a conflict known as Lord Dunmore's War.[19] Chinn insisted that Point Pleasant, preceding as it did the battles of Lexington, Concord, and Bunker (Breed's) Hill by several months, was the first armed engagement of the American Revolution. Kentucky, Chinn went on, did not participate in writing or presenting the Declaration of Independence. The battle of Blue Licks, he further said (Blue Licks was still a part of Virginia at the time, known as Kentucky County, Virginia; today it is in Robertson County, Kentucky), was the *last* battle of the Revolution. It was fought on August 19, 1782, some ten months after Lord Cornwallis had surrendered at Yorktown. A combined force of 350 Loyalists subdued 182 Kentucky militia. Blue Licks, argued many professional historians, was more credible as the last battle than Point Pleasant was as the first. Chinn spoke also of Kentucky's role in gaining ratification of the Constitution in 1788, when twelve residents of the Virginia "Judicial District" participated in Virginia's Convention to vote either for or against a new Constitution. Ten of the fourteen Kentuckians voted *for* the Constitution, an action that did not, of course, deny the Constitution's final approval. Colonel Chinn was so awed by the Constitution that he called it "the greatest document ever penned by human hands."[20]

Most criticisms directed at Colonel Chinn's statements in his last two books about the role of Harrodsburg/Mercer County fell on deaf ears—literally. He had fired relatively few weapons in all his years in the Marine Corps. (His borrowing of the Boone rifle from KHS display and firing it in Shepherdsville was an exception.) Yet he was in daily contact with soldiers and experimental testers who did. The result, after many years, was catastrophic. Personnel at KHS began to notice how loudly and often they had to repeat themselves when they spoke to the colonel.[21] He was quite obviously losing his hearing. Aware of this worsening condition, Colonel Chinn, without mentioning it to anybody, began to prepare for its consequences by learning how to lip-read. Did he ever think of hearing aids? If he did, he never mentioned it to anyone, friends or family.

He told interviewer Ellis in July 1987, "You're talking to a senile old man." And, as an added attraction, "I'm pretty near stone deaf. I was up at Fort Knox yesterday and they gave me a bad report on my hearing." He told Ellis that he did a lot of lip-reading and had been studying this technique since he first became aware of a hearing problem. He couldn't get the "key words" of a conversation. Even if he understood eighteen out of twenty, those two remaining words almost always turned out to be key to the entire conversation.[22] Therefore, he began to request that individuals with whom he conversed stand in front of him and speak slowly and distinctly so that the colonel could lip-read their words. One of the casualties of Chinn's inability to hear well was his use of the telephone. Throughout his careers in the Marine Corps and then at KHS, Colonel Chinn practically had a telephone glued to his ear the whole time. Another concession he made to the aging process was that he had to start wearing thick-lensed eyeglasses, so he could at least halfway see what he was reading.

The citizens of Harrodsburg and surrounding areas loved Colonel Chinn and were loyal, disparaging his detractors and enemies as much as he did. He had lived "multiple lives," and as far as the Bluegrass settlements were concerned, he had excelled in all of them: football player (baseball, too, when called upon), coach, tour guide, restaurateur (in his cave), government sergeant at arms, civil rights leader (except the time he rejected the collection of black newspapers), bodyguard, military man, weapons expert, librarian, author, director and deputy director, KHS, head of military museum in Frankfort, raconteur, and, overall "good old boy." Was there anything George Morgan Chinn *couldn't* do? In the minds of his family, friends, and neighbors, the answer was a resounding *no*. Were there any academics around who could surpass or even match a record like this? The answer again was no. It was clear that Colonel Chinn had led an extraordinarily exceptional life and career (perhaps one should say "careers," since there seemed to be so many of them).

Chinn continued to receive one military award after another (American Defense Preparedness, Bronze Medallion, Legion of Merit, Bronze Star, Navy Commendation, Presidential Unit Citation, Navy Meritorious Service, and "Guest of Honor" at numerous military reunions, such as Veterans' Day) throughout the 1970s and 1980s. He attended a military meeting at the El Toro USMC base in California. The speaker unexpectedly complimented Chinn for his work in aviation ordnance. A few weeks later Chinn responded to the speaker's remarks. "Your recognition of me before such a distinguished gathering . . . was not only deeply appreciated, but I was also thankful the auditorium was dark for the others [in attendance] most certainly would have seen an old Marine sitting there with a lump in his throat the size of a grapefruit."

When he was growing up in and near Harrodsburg, he said, "history was almost like a religion," and it was always the most discussed subject in the area. "History is not a hobby in Harrodsburg," he told one audience after another, "it's a religion."[23] Or at least that was true in earlier days before TV became a major attraction. Nowadays, he grumbled, everybody watched television for six or seven hours a day; history lost out in this change of lifestyle.[24] His books on Harrodsburg and Mercer County, however, brought many residents back to the printed page, a feat of which he was exceedingly proud. He probably should have put a saying of his own creation over his door: "I

Can't Stand Inactivity," a slogan that goes a long way to explain George M. Chinn.

He was constantly asked to compare past centuries with the twentieth. "Every generation seems to have to go through something that scares the 'beejesus' out of them." For Chinn's generation it was poison gas on the battlefield during World War I. It was difficult for people in the late twentieth century to believe that, in addition to poisonous gas, the zeppelins between the two wars "scared the world more than nuclear weapons do today." These words inspired another question from a reporter: What were the early settlers afraid of? "Nothing," was Chinn's quick response. "There wasn't anything that scared them that I can think of, except they were God-fearing people."[25] In 1900, Chinn believed, Harrodsburg's "pioneer past" caught up with the oncoming age of industrialism. It was the greatest symbol in the world when citizen Lafon Riker drove a car down Harrodsburg's Main Street. It seemed that almost everyone in town and surrounding areas came to see this amazing new phenomenon. Chinn wondered: did the observers laugh at it or stand in awe? Did they realize what impact this new technology would have on their lives?[26] And those of their children and grandchildren?

Colonel George Morgan Chinn seems to have entered a melancholy phase of his life in the mid- to the late 1980s. He still worked at his office in downtown Harrodsburg, but he did not give as many speeches as he was accustomed to, or visit KHS and the capitol grounds in Frankfort. He seemed quieter and more retrospective, perhaps summing up his incredibly busy life. His family and friends began to notice he was slowing down in his activities, that he was suffering from breathing problems, and that he had begun to lose weight. Each Sunday in Harrodsburg, he'd take his wife, Cotton, to Sunday school and then drive over to his granddaughter Ruth's house to wait until church services started. Ruth and her family noticed that his breath came with increasing difficulty, certainly justifying a visit to a physician.

His daughter, Ann, and her husband, Howard, took him to see his doctor. While waiting in an anteroom while Chinn had his consultation, they could hear the doctor practically shouting at the colonel (because of his deafness), admonishing him to "get his affairs in order," for he had only a few months to live. Chinn did not mention any of this to Ann and Howard, nor did he tell his wife of sixty-three years anything about it, either. Chinn had earlier undergone surgery to remove shrapnel from his nasal passages.[27] He had endured another painful surgery to remove a cancer from the roof of his mouth. Some attributed his ailments and eventual death to "an apparent heart attack."[28] His granddaughter, Ruth, thought he might be suffering from lung cancer ("ironic for someone who never smoked"—although there are photos showing Colonel Chinn with a good-sized stogie in his mouth).[29]

He died on September 4, 1987 (a Friday), at eighty-five years of age. The patriarch George Morgan Chinn had passed on.

With McClellan Funeral Home officiating, services for Chinn were held on Tuesday, September 8, at the Harrodsburg Presbyterian Church. Several individuals eulogized the colonel. Harrodsburg mayor William Noland said he had "always admired and respected" Chinn for his gun work in World War II and Korea; he was "one of a kind, and the most competent historian [the] county has ever had. . . . you just hate to see his passing." An old friend, Enos Swaine, remembered Chinn as a man of wit and humor: "He had a way of always getting around to the humorous side of a situation. He was a good man. We'll miss him." A tearful Happy Chandler told the gathered mourners that he "had known Chinn since Chinn was a little boy down by the river." He was an "amazing man, one of the toughest persons, mentally and physically, that I have ever known. I'm so shocked, so sad to see him go." Then Chandler divulged, for the first time, why he had abruptly changed his mind so many years ago when Chinn appealed to him for help getting into the military. He had been persuaded by Chinn's argument: "I can't let anyone else do my shooting for me. I've got to do that for myself."[30] Thus began a distinguished military (mostly USMC) career that lasted nearly thirty years.

The burial was performed with full military rites and a rifle military salute (a Marine Honor Guard acted as pallbearers) at Spring Hill Cemetery. Chinn left behind his widow, Cotton, a daughter, Ann, and her husband, Howard Howells Sr., two granddaughters, Ann and Ruth, and a grandson, Howard (Buddy) Howells II.

Some thirteen years later, in 2000, a memorial marker was installed at Harrodsburg on the lawn of the Mercer County Courthouse in honor of the late colonel. Since he had been instrumental in furthering the interests of the KHS Historical Marker program, he would have relished the honor. Several dignitaries were on hand to celebrate his life and accomplishments. Marine Colonel John Marsh told the audience that his unit in Vietnam had used Chinn's grenade launchers against the enemy. Though Chinn was an "internationally recognized weapons expert," he was also known for "his humility, common sense and wit." Kevin Graffagnino, KHS director, praised Chinn's efforts—so many years ago now—to build membership in the society and to create programs to make citizens of the Commonwealth aware of their history. Longtime Chinn friend Frank Sowder spoke of Chinn's love of sports, especially football. He closed his remarks by asserting that "Colonel Chinn was my very good, close friend and I miss him very much."[31] Others on the program of this dedication included Don Dixon of the Kentucky Marine Corps League, Judge-Executive Charlie McGinnis, James Cadell III of the Kentucky Transportation Cabinet, Dianne Wells of KHS, and, of course, members of the family.

The plaque reads, in part:

COLONEL GEORGE MORGAN CHINN, USMC.
This Mercer County native was one of the nation's lead-

ing authorities on automatic weapons. . . . A Marine veteran of WWII and the Korean War, Chinn observed combat use of weapons and served as trouble shooter.[32]

Though the plaque shorted him on his historical services to the Commonwealth, it was nevertheless a fitting tribute to a man whose life spanned almost the entire twentieth century, with all its triumphs and troubles. He had tried his best to extend the former and prevent the latter. And in the contexts of those endeavors, he was, in the best sense of the word, a maverick.

Conclusion

∞

An Assessment of Colonel George Morgan Chinn

The meaning of the word *maverick* is elusive both in definition and connotation. In Peter Mark Roget's *International Thesaurus* (sixth edition, 2001; entry 361.6), there are over two dozen synonyms given for the word. They include *bullethead, pighead, hardnose, bigot, fanatic,* and *purist.* None of these terms quite captures the colonel's character, though another entry, *hardhead,* comes close.

Another entry in Roget's *Thesaurus* (868.3) lists "maverick" characteristics—some hundred of them, in fact: "eccentricity," "originality," "unconformist," "out of bounds," "out of step," and dozens more. George Morgan Chinn inherited enough maverick qualities from his Uncle Kit and grandfather John Pendleton Chinn to come naturally to some of these descriptions. Even at an early age, "eccentricity" was certainly apparent. Some of the stories he wrote for Centre's newspaper, *Cento,* and for the *Harrodsburg Advocate* in the 1920s would have been considered "far out" in the 1960s. "Unconformity" was definitely a Chinn trait. When he worked on military weapons and other devices, he went his own way, even sometimes overruling superior officers in significant decisions. He did this not by rebelling against the government but by using sweet reason on his commanders to show them that on particular projects he was right and they wrong. In most instances, the officers listened well and came to agree with Colonel Chinn. It was well that they did, for Chinn helped to supply weapons to the most advanced and sophisticated armed forces in the world at that time—and, for that matter, throughout history.

He could be called "out of bounds," but only compared with the general society in which he lived. Within his professions (military weaponry, history, and directorship of KHS) he was rarely out of bounds. He may have sworn like a sailor in disapproval of some of the conditions he found in the various military and library facilities he came across, but generally he worked well within the parameters of rules and regulations that had been laid out for each discipline. "Out of step" was definitely a Chinn trait, but again within the *entire* society, not particularly in the disciplines with which he was identified. One of the reasons he so strongly supported the Young Historians was to wean them away from television and other distractions to embrace the study of their country's history and other academic subjects.

All societies could use the services of a Colonel Chinn. This is especially true of democratic societies. It is all too easy, even in a democracy, for citizens to respond to the rantings of demagogues. Colonel Chinn did not invent the phrase "Get a life," but he just as well might have. Live your own life (instead of vicariously through celebrities and "heroes") with your own experiences of happiness, sadness, victory, defeat, rest, tiredness, and all the other things we come to know during our passage through this world. And, above all else, the colonel adjured us never, *never* to automatically accept anyone else's ideas unless and until you have thoroughly analyzed them for yourself. Total conformity, Chinn would assuredly argue, is the worst thing that can happen to a democracy.

Historically, American society has lauded mavericks—if primarily in the abstract—admiring their independence of mind, their determination to speak their own thoughts rather than merely accept somebody else's. However, when a real maverick comes along, the public is first amused and bemused, then frightened by the unorthodox opinions, and finally outraged with the maverick's outlook. Mavericks are never afraid to ask even high-ranking members of government, academia, intelligentsia, and so on the questions they believe need answers.

Did it ever occur to Colonel Chinn that his role in the Marine Corps put him in the category of "high society," not only in the military but throughout the civilian populace as well? If this didn't, his directorship of the Kentucky Historical Society surely did. He loved to lead, primarily by example. As a colonel in the Marine Corps he gave orders sparingly; in fact, he received more orders than he gave, an unusual situation for such a high-ranking officer. No matter if one liked or hated Colonel Chinn, almost everyone agreed that he had been instrumental in helping to win World War II and come to a truce with North Korea. As a library director, he exhibited executive qualities perhaps more than as a weapons expert. Few people were in a position to contradict the things he said and proposed doing. These conditions did not lead to any kind of dictatorial stance; he still, almost always, consulted with lower-ranking personnel at the library, seeking a consensus among them. Only a few times did he suffer a snit, and even more rarely a temper tantrum. There was nearly universal acclamation for his work with KHS. Some said he saved the society from extinction. "It was in better shape when he left it than when he became its director" was a frequently heard expression around Frankfort, the state's capital. Even with all his other accomplishments, however, he frequently told friends and acquaintances that his "highest achievement" was authoring the machine-gun section of the 1958 edition of the *Encyclopaedia Britannica.*[1]

What, then, do we make of Colonel George Morgan Chinn during the early years of the twenty-first century? His talent was a mixture of native intelligence and the pragmatic ability to get things done. Early in his life he had told Captain Longmire that he could take things apart, but had trouble getting them back together. Fortunately, for later combat purposes, he had listened to Longmire and observed; long before he entered the military, he had mastered the ability to reassemble things.

As noted throughout this text, in a significant way, going from bullets to books was a fairly easy transition. One wonders if his exaggerations in reference to certain historical subjects were perhaps deliberate, made to aggravate other historians. It seems clear, for example, that Manifest Destiny would ultimately have occurred with or without the feats of General Clark. Equally clear is that the United States would not have remained a "sliver" of land between the Appalachians and the Atlantic Ocean without General Clark's military activities.

Chinn's value beyond armaments and books lay in his questioning attitude toward most things having to do with government. "Ask the hard questions," Chinn would say, and "keep all levels of government honest." This is an outlook we need today. We need more mavericks of the best kind—the kind like Colonel George Morgan Chinn.

Acknowledgments

Although it has become something of a cliché, it nevertheless remains true that nobody ever writes a book all by him- or herself. An author needs help from numerous sources: personal interviews; telephone calls, both received and made; e-mails by the dozens; collections at big university repositories; and, probably most important, individual librarians who go out of their way to procure sources for the writer. In many instances, the popularity and significance of a subject can be deduced by the number of people who come forward to help with the research. I have many individuals to thank, and I hope I don't overlook any of them. If I inadvertently do, please know that I am most sincerely indebted to you for your help.

At Western Kentucky University, I received help and encouragement from several different sources. History department chair Robert Dietle authorized a travel grant for my researches into Colonel Chinn's life and experiences. I deeply appreciate Robert's support of my publishing endeavors. My graduate assistant in this project is a future historian, and a mighty fine one he will be: Robert (Josh) Howard. He did a superb job of gathering materials and helping me construct various points of view; sometimes he knew what I needed even before I did. WKU graduate and now University of Kentucky (UK) PhD candidate Thomas Lee Anderson was essential to the successful completion of this manuscript, and I thank him most sincerely for his help. My friend and colleague Professor Glenn Lafantasie provided me with some useful information, as did Bowling Green travel writer Gary West. WKU graduate director Professor

Beth Plummer always supported me with a graduate assistant, for which I am appreciative.

The WKU library staff, especially the Interlibrary Loan Department directed by Selma Langford, kept me supplied with reading materials. Thank you. Nancy Richey of the Kentucky Library on WKU's campus was expert in finding useful materials. Nancy has helped me in previous researches, and my gratitude to her runs deep. History department secretary Janet Haynes helped considerably in keeping track of expense accounts and encouraging my work. I started work on this project when former history secretary Marsha Skipworth was on the job; she, too, helped the project along. Former history chair Richard Weigel favored me with a definition of history for this book.

At the University of Kentucky, a young graduate student in history, John Wickre, furnished me with significant materials from the UK Center for Oral Studies and other campus libraries, especially information about Happy Chandler, for which I am grateful. Matt Harris of the UK Libraries helped considerably, and so I thank him. The same is true of Doug Boyd, head of the Louis B. Nunn Oral History Center at UK. Kentucky's most famous horseman, James Bassett, sent interesting materials on how Chinn helped to integrate schools in the 1960s. Tom Appleton of Eastern Kentucky University (EKU) gave some important pointers. EKU emeritus professor of history William Ellis helped me with his interviews of George Morgan Chinn. EKU history secretary Jackie Couture facilitated getting Chinn materials to me. Ron Bryant, director of the Waveland Museum in Lexington, provided interesting episodes in Colonel Chinn's life. Mark Hanna, former executive for the *Herald-Leader*, gave me some interesting observations about Chinn. Nancy Blankenship, Russell Hatter (who explored Chinn's cave), Walter Ford, and all the staff at Mercer County Historical Society deserve my thanks.

Some others who gave much-needed help—particularly

from military perspectives—include Randall Fortson, USN; Alfred V. Houde, USMC; Erin Lombard, USMC; and Fred Beatty, USAF; as well as Matthew Jackson. The author also thanks Daniel Jackson for taking care of the computer glitches, which always seemed to occur at the most crucial of times. My old friend Don Stringer, now of Williamsburg, Virginia, encouraged this project as he did others in the past. Bill Pratt gave some useful information on military weaponry, for which I am grateful. I thank Mr. Samuel (Dutch) Hillenburg for sharing some stories with me about Colonel Chinn.

In Louisville, many people kindly responded to my requests for assistance. Librarian Kathie Johnson-Burger was helpful. At the Louisville Free Public Library, Joe Hardesty responded to my request for materials with alacrity, as did Tom Owen at the University of Louisville, as well as Carrie Daniels. At the Filson Society, Jim Holmberg was, as always, friendly and helpful, as was Cassie Bratcher.

My friend James Klotter in the History department at Georgetown College gave me some interesting insights into Colonel Chinn. Nicky Hughes and Susan Hughes are quite knowledgeable about Colonel Chinn, especially concerning his tenure as director of the Kentucky Historical Society. Russell Harris is a mainstay at the society, having produced many articles and treatises on Kentucky history makers. He sent several informative e-mails and other materials to me that facilitated my research on Chinn. Other individuals at the society who helped me include Jennifer DuPlaga, Dr. Darrell Meadows, Sara Milligan, and Bill Bright. Down the road a ways from Frankfort is Danville, Kentucky, home of Centre College. The librarian there, Bob Glass, copied Chinn material from the college's archives and sent it to me, for which I thank him. I thank Donald Stringer of Williamsburg, Virginia, for helping me.

The Howells family was most helpful in supplying me with materials and stories about their grandfather. Buddy, the grandson, and Ann and Ruth, the granddaughters, almost always referred to Chinn by his first name, George. I am happy to report that none of the Howells family tried to influence how this book was to be written. I thank them all for their help. Byron Crawford, a former journalist for the *Courier-Journal*, sent many valuable e-mails while this book was being researched and written, as did Kandie Adkinson of Frankfort.

Professor Drew Harrington of Cumming, Georgia, helped with a definition of history. Professor Robert Norrell of the University of Tennessee, Knoxville, did likewise. Jacquelin Sims of Catawba College in North Carolina sent materials dealing with Chinn's coaching days at that institution.

Other individuals helpful in one way or another were David Bettez, Kate Black, Cassie Bratcher, Susan Brown, Keith Bryson, Mac Coffman, Jennifer Cole, B. J. Gooch, Jamie Helle of Boyle County Library; George Kontis, John McGee, Carolyn Patterson of the Mercer County Library, Ben Samuels of the Harvard newspaper staff, Scott Schurz, and Jack Supplee. Anna Armstrong helped put several photos in good shape for use in this book. I am grateful for her help.

If I have overlooked anyone, please forgive me. I appreciate your help and advice.

I must thank my growing family for their love and support: my beautiful daughters, Beverly and Hilary, and my fine sons, Daniel and Matthew. I hope I can get all the grandchildren lined up properly: granddaughters Colleen, Megan, Katie, Gwennyn, and Ciara, and grandsons Travis, Patrick, Austin, Liam, Rowan, Henry, Isaac, Oliver, and Cranley. And the "grands": Cora, David, Finn, and Kellan. Also Steve, Ling, Arthur, and Elaine. And, as always, first and foremost: Pat. Bless you all.

A Note from the Publisher

Carlton Jackson completed the manuscript for this book, submitted it for publication, and read and responded to two peer reviews before he passed away on February 10, 2014. Unfortunately, he was unable to participate in the subsequent copyediting and proofreading that routinely occur in the publishing process. The University Press of Kentucky made every effort to remain true to Dr. Jackson's words and to seek the advice of his family members and associates when clarification was required. Matthew Jackson and Daniel Jackson, Carlton Jackson's sons, reviewed the copyedited draft of his manuscript and worked with the press to get this book to print. Although the book carries a posthumous publication date, the press is fortunate to have received a completed draft from the author and to have benefited from his guidance at several critical junctures in the publication process.

Appendix A

∾

Chinn's Education, Awards, and Achievements

Schooling

Millersburg Military Institute, Millersburg, KY, 1915–1920
Centre College, Danville, KY, 1920–1924
U.S. Marine Corps, December 9, 1941, commissioned as a
  second lieutenant
USMC Aviation Ordnance School, Quantico, VA
Michigan State University, Lansing, Ordnance
University of Cincinnati, Gage Laboratory
Instructor, USMC Aviation Ordnance School, Quantico, VA

Military Awards

World War I
Victory Medal from Millersburg Military Academy

World War II and Korea
Legion of Merit, Bronze Star (Korea), Navy Commendation
Medal with four bronze stars, Presidential Unit Citation, Navy
Unit Citation, Organized Reserve Medal, American Defense
Medal, WWII Victory Medal, American Campaign Medal,
Navy Occupation Medal (Japan), Korean Service Medal, United

Nations Service Medal (Korea), Korean Presidential United Citation, U.S. Marine Corps Presidential Naval Unit Citation, five letters of personal commendation: three by secretary of the navy, one by USMC commanding general, Pacific Area, one by commandant, USMC

Vietnam

Called back to active duty, 1966. Designed and tested MK-19, MK-20, and others. Received Navy Meritorious Public Service Citation by order of secretary of navy for contributions to Vietnam war effort with "state of the art" weapon design. This decoration is the highest the naval service can bestow on an officer in a retired status

Inventions and Patents

U.S. Navy's EX-6 20 mm aircraft machine gun
U.S. Navy high-velocity 40 mm MK-19 machine gun
U.S. Navy low-velocity 40 mm MK-20 machine gun
MK-22 high-velocity 20/30 mm aircraft machine gun
Chamber lubricator for 20 mm Hispano-Suiza machine gun
Blow-back adapter for .50 caliber M-3 high rate of fire aircraft machine gun
Extractor depresser for .50 caliber M-2 basic machine gun
Flameout eliminator for gun installation F7F fighter aircraft
Slip chamber for 20/30 mm aircraft cannon
Muzzle blast combustion control device for 20/30 mm aircraft cannon
Belt pull accelerator for wing-mounted .50 caliber M-2 machine gun
Continuous-flow ammunition cans for turret mounting aircraft M-2 machine guns

Appendix B

~

Chinn's Publications

Encyclopedia of American Hand Arms. Huntington, WV: Standard, 1940.

The History of Harrodsburg and the Great Settlement Area. Harrodsburg: self-published, 1985.

Kentucky: Settlement to Statehood, 1750-1800. Frankfort: Kentucky State Historical Society, 1975.

The Machine Gun. Vol. 1, *History of the Evolution and Development of the World's Machine Guns.* Washington, DC: U.S. Navy, 1949.

The Machine Gun. Vol. 2, *The Development of Automatic Weapons prior to and during World War II by the Soviet Union and Its Satellite Countries.* Washington, DC: U.S. Navy, 1950.

The Machine Gun. Vol. 3, *Cannon and Machine Gun Development. . . . Beginning with WWI and through the Korean Conflict.* Washington, DC: U.S. Navy, 1953.

The Machine Gun. Vol. 4, *Theory and Basic Design of All Automatic Weapon Systems.* Washington, DC: U.S. Navy, 1955.

The Machine Gun. Vol. 5, *Engineering Designs Development, 1952–1987.* Ann Arbor, MI: Edwards, 1987.

Through Two Hundred Years—Pictorial History of Mercer County, Kentucky. Harrodsburg: self-published, 1974.

Tools of Pioneer Warfare. Lexington: Herald, 1938.

Notes

Introduction

1. Chinn, interview by Ellis.
2. This is exactly why the armed forces would not give him a final discharge. He knew too much and was too valuable. He stayed in the Marine Corps for twenty-six years, and only a few times during that period did he weigh under three hundred pounds. Though military physicians questioned him several times about his weight, they quickly learned that recommending diets for Chinn was perilous to their own well-being, given the role the colonel played in martial affairs. Some military physicians came to believe that Colonel Chinn was a "plant" to keep the doctors on their toes. Chinn was so large, and if he were walking across the base, some doctors thought it was their mission to interrogate him about his weight. When they learned who he was, of course, they backed off. Howells, interview by author February 7, 2012.

Colonel Chinn had both classified and unclassified information before him; too much for the government easily to dismiss him from the armed forces. He finally did get out of the military (first in 1959, only to be recalled in a "consulting" role for Vietnam in the 1960s). Thus, he earned—among other decorations—the Legion of Merit (WWII), Bronze Star (Korea), and Meritorious Service (Vietnam). Norman, "The Kentucky Colonel of Pen and Sword," 17.
3. Chinn, "Goes Hunting."
4. Anderson, "Colonel George Morgan Chinn," 197.
5. Chinn, *Through Two Hundred Years*, 1.
6. Chinn, interview by Crawford.
7. Ibid.
8. Ibid. See also Norman, "The Kentucky Colonel," 16.
9. Chinn, interview by Crawford. See also Kontis, "'Buddy' Howells," 96.
10. *Communicadet* (1920), 2.
11. Kontis, "'Buddy' Howells," 86. Millersburg closed its doors long after Chinn attended. It was converted to a prep school. Howells, interview by author, February 7, 2012.

12. Colonel Best founded Millersburg Training School in 1893, and the school was soon renamed Millersburg Military Institute. It went from having 1 boarder in 1893 to 93 in 1920. By 1920, some 1,300 to 1,400 students had been enrolled at Millersburg Military Institute; 219 were graduated, of whom 80 percent ultimately received academic degrees from universities and colleges throughout the United States. In World War I over two hundred former cadets enrolled in the U.S. Army, between 60 and 70 of them commissioned officers. George M. Chinn, one of 9 cadets, received his diploma from the academy. Commencement program, June 4, 1920, Howells Family Collection.

13. Chinn, *The History of Harrodsburg*, 2.

14. Chinn, interview by Ellis.

15. Bright, "Mercer County's Modest Marine."

16. Norman, "The Kentucky Colonel," 16.

17. Carr, "Mercer County."

18. Martin, "Gun Experts Hail Kentuckian's Aid."

19. Bright, "Mercer County's Modest Marine."

20. "Prisoners Brutally Punished."

21. "Due to Tuberculosis."

22. "News and Roundabout."

23. "Colonel Jack Chinn Seems to Be on the Warpath."

24. According to *Wikipedia: The Free Encyclopedia*.

25. Chinn, *History of Harrodsburg*, 132. See also *Wikipedia: The Free Encyclopedia*.

26. *Seattle Times Sports News*, May 3, 1966.

27. Ibid.

28. *Park City Daily News* (Bowling Green, KY), September 2, 2012.

29. *Seattle Times Sports News*, May 3, 1966.

30. *Wikipedia: The Free Encyclopedia*.

31. Pearce, "Col. Jack Chinn."

32. Ellis, "When Col. Jack Chinn Got 'Het Up.'"

33. Ibid.

34. Ann Howells, e-mail to author.

35. Chinn, interview by Crawford.

36. Ironically, one of George M. Chinn's associates at the Kentucky Historical Society in the 1980s was William Buster, whose great-uncle, Ephraim Lillard, stood beside J. P. Chinn when the shooting started. See Zuercher, "That Day Frankfort Turned Dodge City."

37. Chinn, interview by Crawford.

38. See *Harrodsburg Herald*, April 3, 1903, reprinted in Bailey, *Murders, Mischief, Mayhem*, 5–6.

39. Chinn, interview by Crawford.

40. Miller, "Historically Speaking."

41. Chinn, *History of Harrodsburg*, 115–16.

42. Town council ordinance, Harrodsburg, 1860, forbidding whites and blacks to be buried in the same cemeteries.

43. Norman, "The Kentucky Colonel," 17.

44. See Hanna, letter to the editor.

45. Norman, "The Kentucky Colonel," 17.

1. What's in a Name?

1. Howells, interview by author, September 12, 2012.

2. Norman, "The Kentucky Colonel," 17.

3. Chinn, interview by Ellis.

4. Ibid.

5. Kontis and Chinn became close friends after the latter asked for a copy of a report that Kontis gave to an audience at the Advance Engineering Group in Louisville. "He wanted a copy of everything. Thank God because nobody else was keeping a copy." Kontis, e-mail to author.

6. Chinn, interview by Ellis.

7. Ibid.

8. Ibid.

9. Crawford, e-mail to author, February 8, 2012.

10. Gabler, "George Chinn." See also Ziegler, "C6-H0."

11. Robertson, *The Wonder Team*, 273.

12. Ibid., 325.

13. Ibid. The author thanks Dr. Robert W. Robertson Jr. for his help and encouragement on this biography of Colonel George Morgan Chinn.

14. Norman, "The Kentucky Colonel," 17.

15. Bright, "Mercer County's Modest Marine."

16. Norman, "The Kentucky Colonel," 17.

17. Robertson, *The Wonder Team*, 122. See also "Teams in Good Condition."

18. Ziegler, "C6-H0."

19. Chinn, "Goes Hunting."

20. Robertson, *The Wonder Team*, 209–10. Also see Ziegler, "C6-H0."

21. Ziegler, "C6-H0," 2.

22. Brown, *The Legend of the Praying Colonels*, 58–59.

23. Thurman, *Run That by Me Again*.

24. Ibid.; Brown, *The Legend of the Praying Colonels*, 50.

25. Ziegler, "C6-H0," 6.

26. Ibid., 5.

27. Daly, "Center for Centre."

28. *Harvard Alumni Bulletin*, 117.

29. Ziegler, "C6-H0," 4.

30. Ibid. See also "David Skunks Goliath at Harvard"; Samuels, "Remembering a Forgotten Upset," 1–4; Robertson, *The Wonder Team*, 252–73; Thurman, *Run That by Me Again*.

31. *Harvard Alumni Bulletin*, 115.

32. Ziegler, "C6-H0," 7.

33. Quoted in Brown, *The Legend of the Praying Colonels*, 56.

34. Ibid. Cancellation of Centre's classes was ironic, to say the least, especially since many of the college's professors had not approved student absences to attend the game.

35. Robertson, *The Wonder Years*, 195.

36. Ziegler, "C6-H0," 4.

37. Samuels, "Remembering a Forgotten Upset."

38. Samuels, e-mail to author.

39. Ibid.

40. Samuels, "Remembering a Forgotten Upset," 3.

41. Thurman, *Run That by Me Again*.

42. Ibid., quoting Hope Hudgins.

43. Robertson, *The Wonder Team*, 152. Bolstering Moran's "absolutely necessary" statement was the fact that Centre lost only two games in 1922: one to Harvard, 24–10, and the other to Auburn, 6–0. The Praying Colonels ripped through Ole Miss, 55–0; Clemson, 21–0; Louisville, 32–7; South Carolina, 42–0; and Washington & Lee, 27–6. From Moran's point of view, the team was doing so well that he didn't need Chinn's services, a point with which Chinn vehemently disagreed.

44. All references to the Shelbyville-Pleasureville football game come from Crawford, "Henry Ran Ringers around Shelby." The author wishes to express his gratitude to Byron Crawford for his support of this biography of Colonel Chinn.

45. Chinn, "Goes Hunting."

2. Football and Caves

1. In the 1920s, it was apparently a "coming-of-age" routine among young ruffians to see how many cars and their passengers they could "stone," especially in rural areas where the horse was still, many times, regarded as the chief form of transportation, and animus against automobiles was still felt. This practice was in all likelihood a prelude to the "mailbox bashing" of many years later; both were pitiful attempts to show that youngsters had reached "manhood."

2. "Rock Thrower"; Howells, interview by author, May 21–22.

3. Robertson, *The Wonder Team*, 502.

4. This organization may or may not have been the soup company. Although Campbell soup goes back to 1869, the author has found no evidence that Campbell ran the company for which Chinn worked at this particular time.

5. Robertson, *The Wonder Team*, 523.

6. Ibid.

7. "Chinn Describes Prexy's Defense."

8. However, a former senior officer of the KHS remembered a manuscript biography of Moran (never published) that alleged that Moran *did* pay players, or, if not Moran, someone else on the Danville campus did. Anonymous e-mail to author.

9. Howells, interview by author, September 16, 2012.

10. Chinn, interview by Ellis.

11. Kontis, "'Buddy' Howells," 86. See also Howells, interview by *Small Arms Defense Journal*, 97.

12. *Catawba College Yearbook, 1931*, 115. See also *Pioneer* (Catawba's newspaper), October 20, 1930.

13. *Pioneer*, October 20, 1930, 1.

14. Ibid.

15. *Pioneer*, December 19, 1930, 1.

16. Howells, interview by author, September 25, 2012.

17. Ibid.

18. Chinn, letter to Chandler, November 17, 1931, University of Kentucky Libraries.

19. Ibid.

20. For references to this event, consult Pearce, *The Colonel*, 50–53. The local paper, the *Times-Tribune*, spelled Sanders's name in its report as Saunders. *Corbin (KY) Times-Tribune*, May 8, 1931; Howells, interview by author, September 2012. The problem occurred long before "Colonel" Sanders became famous for his fried chicken. He ran a gas

station, and his billboard advertising his business helped divert traffic away from Matthew Stewart's Gulf station. Stewart constantly painted over the sign, only to have Sanders restore it again. In the shooting, a representative from Shell Oil, Robert Gibson, was shot and killed, presumably by Stewart.

21. Crawford, "Col. Chinn's Life."

22. See Okrent, *Last Call.*

23. Crawford, "Col. Chinn's Life"; Howells, interview by author, September 17, 2012.

24. *Harrodsburg Herald,* July 13, 2000.

25. http://www.topix.com/forum/city/harrodsburg-ky. TD5SFMJODMKT.

26. Kontis, "'Buddy' Howells," 97.

27. Crawford, "Col. Chinn's Life."

28. Ibid.

29. Howells, interview by author, September 19, 2012.

30. Chinn, interview by Ellis.

31. Norman, "The Kentucky Colonel."

32. "Two Wounded in Shooting Affair near Brooklyn Bridge on Monday," *Harrodsburg Democrat,* May 13, 1930.

33. Kontis, "'Buddy' Howells"; Howells, interview by author, September 17, 2012.

34. Kontis, "'Buddy' Howells."

35. Hanna, letter to author; Howells, interview by author, September 16, 2012.

36. http://www.topix.com/forum/city/harrodsburg-ky TD5SFMJODMKT.

37. Crawford, "Kentucky's 'Buddha of Ballistics.'"

38. Kontis, "'Buddy' Howells," 94.

3. Odds and Ends; or, Here and There

1. Martin, "Gun Gadget Wins Fame for Chinn." It is true that the Chinns could have moved back into their house in downtown Harrodsburg. It was one of the oldest in Mercer County and was constructed by adjoining two log houses. George and Cotton were married in the living room of that house. In the twenty-first century, grandson Buddy Howells and his own son are restoring the Chinn residence "room by room." See Kontis, "'Buddy' Howells," 94.

2. A "Tyrone" ferry apparently came originally from Tyrone County, Ireland. Since ferries were used to transport cars, trucks, wagons (along with the horses and mules that pulled them), and pedes-

trians, they were usually of great dimensions. It took a considerable amount of time for George and Cotton, frequently hiring workers, to convert the ferry into living quarters.

3. Howells, interview (telephone) with author, February 22, 2013.

4. George M. Chinn to A. B. "Happy" Chandler; November 27, 1931, box 3, Chandler Special Collections, University of Kentucky Libraries. Noticeably, even at this early date of their friendship, Chinn's salutation to the lieutenant governor was "Dear Happy."

5. Ibid.

6. Ibid.

7. Chandler, letter to Chinn, University of Kentucky Libraries.

8. The author thanks John McKee of the Kentucky Legislative Research Council for this information. E-mail to author. Chinn served as assistant sergeant at arms for the state senate from January 1, 1932, to January 1, 1936, and from January 1, 1940, to January 1, 1942. Unfortunately for Chinn, the position did not qualify him to collect an employee retirement contribution. He did serve as sergeant at arms from January 1, 1936, to December 31, 1937. Later, he claimed to have been the sergeant at arms from January 1, 1940, to January 1, 1942. The Kentucky Employees Retirement System found this information to be incorrect. Records indicate that Porter Tanner was the sergeant at arms from January 1940 to 1942. Chilton, executive secretary of Kentucky Employees Retirement System, letter to Chinn, Howells Family Collection.

9. Kontis, "'Buddy' Howells," 88. Chinn, letter to Chandler, November 27, 1931, University of Kentucky Libraries; Chinn, letter to Chandler, November 10, 1937, University of Kentucky Libraries; Martin, "Gun Gadget Wins Fame for Chinn"; "Lt. Col. Chinn is One of the Top Men in Ordnance."

10. Martin, "Gun Gadget Wins Fame for Chinn."

11. Norman, "Kentucky Colonel of Pen and Sword."

12. Chandler and Trimble, *Heroes, Plain Folks, and Skunks*, 118. When former heavyweight boxing champion James Dempsey came to Kentucky, Chinn was hired, on Chandler's recommendation, as his bodyguard. Just why a heavyweight boxer would need a bodyguard was never discussed.

13. Ibid.

14. Howells, interview by author, April 11, 2012.

15. Chinn was away so often seeing to his duties that once when he did knock on the front door on a visit home, his little daughter, Ann,

did not recognize him. She ran for her mother, yelling, "Mommy, that man's here again!" Bright, "Mercer County's Modest Marine."

16. Thierman, "Modern Quarrying"; Norman, "The Kentucky Colonel of Pen and Sword." There is a photo of the Chinn house atop the palisades overlooking the Kentucky River in *Kentucky Ancestors*, April 1976.

17. Thierman, "Modern Quarrying."

18. Pearce, "Col. Jack Chinn."

19. Chinn house specification sheet, prepared by Mary Chambliss, Mercer County, Kentucky, Library.

20. Kontis, "'Buddy' Howells," 94.

21. Pearce, "Col. Jack Chinn." In later years, "ghostbusters" in the area found the Chinn residence and, despite warnings from the heirs, went through the place looking for inhabitants of past times. Are there such things as ghosts? Many people on the Internet seem to think so. See *Haunted Houses in Mercer County*, http://www.topix.com/forum/city/harrodsburg-ky / TCAA3C8R7GOLH.

22. Roberts, "Chinn's Inn." The desire for good-quality whiskey seemed to run in the Chinn family. One of Colonel Chinn's distant relatives, John G. Carlisle, was at various times Speaker of the U.S. House of Representatives, a senator from Kentucky, and a treasury secretary. When Colonel Chinn was in Washington, DC, he read through Carlisle's papers at the National Archives and became an authority on bourbon, although he had sworn off it years before (see ibid., 14). Carlisle wrote the "Pharmacopoeia Law," creating guidelines for distilling and bottling whiskey. He also wrote the "Carlisle Allowance," the formula for taxing the grain from which whiskey is made. Carlisle himself was a heavy drinker but could become instantly sober whenever there was important work to do. On one-hundred-proof bourbon, the stamp affixed over the cap has a portrait on it: that of John Griffin Carlisle. His memory (he died in 1910) was frequently evoked during the period of making the Rochemont film. See Hicks, "Before You Twist Cap."

23. Chinn, letter to Chandler, November 20, 1937, University of Kentucky Libraries.

24. Ibid.

25. Chandler and Trimble, *Heroes, Plain Folks, and Skunks*, 118.

26. The author is indebted to Bowling Green, Kentucky, travel writer Gary West for this information.

27. Dedman, interview by West.

28. Ibid.

29. Kontis, "'Buddy' Howells," 89.

30. Martin, "Gun Experts Hail Kentuckian's Aid."

31. Ibid. Interestingly enough, it was a Colonel Best who ruled over the fortunes and activities of the cadets, including George Morgan Chinn, at Millersburg Military Institute (see chapter 1). If there was ever any connection between the two Colonel Bests, such was not mentioned in any of the military accounts.

32. Bright, "Mercer County's Modest Marine."

33. Ibid.

4. Semper Fi

1. Hughes, interview by author.

2. Howells, interview by author, September 18, 2012.

3. Norman, "The Kentucky Colonel of Pen and Sword," 18.

4. Hackett, "Boone's Rifle Embarrassed Weapons Expert."

5. "It's No Surprise that Kentucky Rates Third Growing Marijuana." In 2014, a battle is raging in the Kentucky state legislature once again over whether to legalize "industrial hemp" under conditions of closely monitored cultivation. In the nineteenth century, Kentucky was the leading grower of hemp, which, according to Chinn, imparted a strong "euphoric" effect.

6. Kentucky, even with a reputation of hemp as its "top" crop, was not alone in being authorized by the Department of Agriculture in World War II to "commission" the cultivation and production of hemp and "other materials crucial for producing marine cordage, parachutes and other military necessities." The Department of Agriculture's program Hemp for Victory tried to convince U.S. farmers to "plant hemp by giving out seeds and granting draft deferments to those who would stay home and grow hemp." Altogether, American farmers registered in Hemp for Victory harvested some 350,000 acres of hemp during the war. See www/pbs.org/wgbn/pagers/frontline/shows/dope/etc/cron.html. There is some discussion in the Chinn family today (2014) about whether the "Chinn" referred to here is the colonel or his father, George M. Chinn Sr. The latter had a contract with the government for industrial hemp back in the 1930s; press reports on Chinn in 1983 cite him as director emeritus of the Kentucky Historical Society. That could only have been Chinn Jr.

7. Chinn, interview by Newland.

8. Hicks, "Marijuana's Roots."

9. "Lt. Col. Chinn is One of the Top Men in Ordnance." *Windsock,*

in which this article appeared, was published every Friday "as an activity of the Special Services Department."

10. Ibid. Of course, Frigidaire was not alone in retooling its equipment from a domestic to a wartime production. General Electric did so throughout the country, as did Oldsmobile in Michigan. From time to time Chinn was asked to advise and consult with the companies working for the war efforts.

11. Ibid.

12. Printed in Chinn, *The Machine Gun*, 5:xiii.

13. Martin, "Gun Gadget Wins Fame for Chinn." Many of the individuals who came to know this story thought it ironic that a peaceful clergyman helped to put a weapons expert into what became a high-ranking job in the military. If Chinn ever imagined such a scenario, he stayed quiet about it.

14. He spent more time during the war at Patuxent than at any other base.

15. "Lt. Col. Chinn is One of the Top Men in Ordnance."

16. Howells, interview by author, September 18, 2013. "Fat Boy" was, of course, one of the names given to the atomic bomb.

17. Martin, "Gun Gadget Wins Fame for Chinn."

18. Brown, letter to Chinn, Howells Family Collection.

19. Forrestal, letter to Chinn (probably 1946), Howells Family Collection.

20. Hoover, letter to Schoeffel, Howells Family Collection.

21. No doubt Chinn's admirers would have been surprised to learn that, in addition to being an expert in heavy military ordnance, he also wrote poems. A sample of Chinn's poetry that inspired "nostalgia" from Millersburg Military Institute, with apologies to Edgar Allan Poe, was:

Twas on a December night, bleak and dreary
That I dreamed on Little Mary, whom I adore
Twas then I heard a rapping
As if someone gently tapping,
Twas only John's shirt tail flapping
Only this and nothing more.

And a few stanzas later, after a considerable snowfall, Chinn faced the ordeal of leaving his warm bed:

At last the bugle blew
From under the cover my carcass I drew
While straightway into my breeches I flew
And from under the snow my socks I drew.

He was sixteen years old when he wrote this "poem." It was not published until some time after he wrote it. Untitled poem, *Centre College Cento*, November 21, 1923.

22. Gleason, review of *The Machine Gun*, 219–20.

23. Chinn et al., *The Machine Gun*, 2:3. Even though Chinn usually avoided small arms (except for a private practice known as "plinking") in favor of large military weapons, he and fellow Kentuckian Bayless Evans Hardin produced a tome entitled *Encyclopedia of American Hand Arms*. R. K. Lewis reviewed it for *Army Ordnance*, praising it for the color illustrations accompanying each of the 382 pieces in the book. "Here is a classification of American short arms from their beginning in the early eighteenth century through to the present time."

24. The Smith and Wesson pistol became the "Russian revolver" (Chinn et al., *The Machine Gun*, 2:5).

25. Crawford, e-mail to author, December 8, 2011.

26. His World War II and Korean inventions, restorations, and modifications included the "Blizzard Buggy," which created synchronization of four twenty-millimeter cannon "triggered as a single unit," which helped against kamikaze strikes; a "flameout" eliminator; a thirty-millimeter gun called EX-6; a chamber lubricator that modified the high-speed twenty-millimeter Hispano-Suiza for deck mounting on "riverine" patrol boats; the MK-19 forty-millimeter high-velocity grenade launcher; and the MK-22 dual-caliber cannon. This list is excerpted from a speech given November 11 (no year or place listed) by General Pat Patrick in honor of the Marine Corps birthday.

27. Wood, letter to Chinn, Howells Family Collection.

28. Ramsey, "Ex-Marine Heads Historical Society."

29. Flowers, interview by Brinson, January 24, 2002, Oral History Project, Kentucky Historical Society.

30. Pyle, "Modernizing Our Small Arms," 21, 22.

31. Ibid., 13.

32. *Wikipedia: The Free Encyclopedia.*

33. Stringer, e-mail to author.

34. Anderson, e-mail to author, November 5, 2013.

35. Ibid.; *Wikipedia: The Free Encyclopedia.*

36. Howells, interview by author, July 22, 2013.

37. Norman, "The Kentucky Colonel of Pen and Sword."

38. Anderson, e-mail to author, April 8, 2013.

39. The author is indebted to Travis Welch and Matthew Jackson for these descriptions.

40. Howells, interview by author, September 17, 2013. Many years after the 1950s and 1960s, the United States joined with the United Nations peacekeeping efforts in Somalia. In October 1993, Somali forces surrounded the capital, Mogadishu. Captain Charles F. Ferry and Company A of the Fourteenth Infantry were sent to clear them out. "The fire fights during the two days," he reported, "demonstrated capabilities of the MK 19." He went on about the M-19: "It's big and loud and the bad guys respect it." See Ezell, *National Defense*, September 1995, 27.

41. Anderson, interview by author.

42. Glenn, *John Glenn*, 215.

43. Bergin, letter to Chinn, Howells Family Collection. See also *Kentucky New Era*, January 9, 1969.

44. Bergin, letter to Chinn.

45. Glenn, *John Glenn*, 215.

46. Ibid.

47. Ibid.

48. Ibid., 214.

49. Patrick, speech in honor of the Marine Corps birthday.

50. Geiger, letter, Howells Family Collection.

51. See Jackson, *Zane Grey*, 144–45. For information on Billy Vaughn, see Jackson, *P.S. I Love You*, 14–15.

52. Kontis, "'Buddy' Howells," 94.

5. "Whose History Is It, Anyway?"

1. Again, the author is indebted to Bowling Green travel writer Gary West for this information.

2. Marine birthday speech by Colonel George M. Chinn, USMC, no year or location given, Howells Family Collection. The speech was probably delivered at Beaumont Inn in Harrodsburg. This became a set speech for all the others he gave, none of which indicate, unfortunately, the year or location. Since the Marine Corps's birthday is in November, these speeches were usually given around the middle of the month. It was widely aired throughout the state that Happy Chandler also used this story in his speeches. Unfortunately, it does not appear in any of his writings.

3. Ibid.

4. Ibid.

5. Col. George M. Chinn, notes for Mother's Day speech, May 30, 1956, Howells Family Collection.

6. Brown, "The Colonel Stirred Controversy."

7. Harris, e-mail to author, May 7, 2013.

8. Ibid.

9. Klotter, interview by author.

10. Harris, e-mail to author, October 8, 2012.

11. Magruder letter regarding Chinn, November 2, 1958, Marine Corps Museum.

12. Gleason, review of *The Machine Gun*, 220.

13. Chinn, interview by Ellis.

14. Clark, interview by Marshall.

15. Norman, "The Kentucky Colonel." The snide nature of Chinn's remarks was manifest.

16. Day, "Mercer County."

17. Hartley, "Has Guns, Still Traveling." Unfortunately, there do not seem to have been any recordings or even transcriptions of this hair pulling.

18. Klotter, "Thoughts on Colonel Chinn" (statement to author).

19. Chinn, letter to Clift, Louisville Public Library.

20. Manning, notes on Colonel Chinn's reaction to the "Communique" matter, Louisville Public Library.

21. Hanna, letter to author.

22. Bryant, e-mail to author, March 9, 2013.

23. Hanna, letter to author.

24. Bryant, e-mail to author, February 13, 2012.

25. Flowers, interview by Brinson, January 24, 2002, Oral History Project, Kentucky Historical Society.

26. Blankenship, interview by author.

27. Bryant, e-mail to author, March 9, 2013.

28. *American Historical Review* (June 1989).

29. Hanna, letter to author.

30. Harrington (professor of history, emeritus, Cumming, GA), e-mail to author.

31. Norrell (professor of history, University of Tennessee, Knoxville), e-mail to author.

32. "Amateur Historians," anonymous untitled essay, n.d., in author's possession.

33. Weigel (professor of history, Western Kentucky University), e-mail to author.

34. Achenbach, "Gettysburg."

35. Norman, "The Kentucky Colonel." Another prominent histo-

rian, explaining that Chinn almost always took "stands" on matters, said, "Historians ought not to take stands." The statement itself, of course, is a stand.

36. Earlier examples of formal and informal explanations of U.S. history were Zane Grey and Frederick Jackson Turner. Grey wrote for the greater public, Turner for intellectual readers. There is no doubt about who was the more widely known of the two: it was, of course, Zane Grey, by a wide margin. See Jackson, *Zane Grey*, 124–25.

37. Chinn calculated that he would be away in Virginia and Vietnam for six to eight months. He wanted to appoint Frank Sower to act as KHS director in his absence. Though touched by the offer, Sower declined. He happened at that time to be the mayor of Frankfort. Sower, interview by Brinson, January 4, 2002, Oral History Project, Kentucky Historical Society.

38. Magruder to Chinn, April 28, 1961, Marine Corps Museum. Chinn's response to these notions was "Frankly, I'm flattered they [apparently the USMC] think there might be some mileage left in an old tread like me." "Kicking Up the Dirt."

39. Magruder to Chinn.

40. Gaddis, interview by Grider, September 12, 1990, Oral History Project, Kentucky Historical Society.

6. Back to the Cave

1. Russell Hatter kept a detailed diary from January 2, 1966, to November 3, 1968, of his and Clifford's activities in George and Cotton's cave. They noted carefully how much mud they removed, water flow, seeping problems, and locations of springs. The author is indebted to Mr. Russell Hatter for allowing me to study and use these descriptions of the Chinn cave. Much of this chapter is based on Hatter's diary, with Mr. Hatter's permission.

2. Nevertheless, a huge job lay in front of them. Hatter Diary, January 1966–November 1968.

3. Hatter, e-mail to author.

4. Hatter Diary, January 1966–November 1968.

5. Ibid.

6. Ibid.

7. Ibid.

8. Ibid. Modestly, Hatter in 2013 denied being a "professional" spelunker except in the mind of Colonel Chinn himself. At the time in

1966, Chinn thought that he was "just a few steps away" from a fabulous fortune to be found in the dark interiors of his cave.

9. Hatter Diary, January 1966–November 1968.

10. Ibid.

11. Ibid.

12. Hatter, e-mail to author. Everything is still fairly much today as it was forty-five years ago. And who knows? One or more of the rooms Hatter and Clifford "heard" may very well be full of stalactites, stalagmites, and other exotic relics. Colonel Chinn seems not to have shown any great interest in tourist attractions from 1968 to the time of his death in 1987. In the 1980s, for a short time, he did allow truck drivers taking materials for the renovations and repairs of the nearby Dix Dam to stop at the cave's entrance and eat their dinners (the noontime meal). This practice had to be discontinued, however, because of near accidents caused by curves, narrow lanes, and high drop-offs at that point on Highway 68. Also, big semis were literally tearing up the road going to and fro between Harrodsburg and Lexington. The county government (Mercer) finally stopped semis from using the route altogether. "Cave on 68," online comments, November 2011.

13. Ignatius (secretary of the navy), letter to Chinn, Marine Corps Museum. The author sincerely thanks Alfred Houde and his staff at the Marine Corps Museum at Quantico, Virginia, for sending valuable materials to me dealing with Colonel Chinn.

14. Matter, letter to commandant, Marine Corps, Howells Family Collection.

15. Chinn, letter to Chapman, Marine Corps Museum.

16. Chapman, letter to Chinn, Marine Corps Museum.

17. Hartley, "Has Guns, Still Traveling." The author could not find any documentation that indicated Chinn sought professional medical help, either military or civilian, to relieve the stress. It is possible that he did not know he suffered from stress.

18. "A Big Man Becomes Bigger."

19. Day, "Mercer County."

20. Chinn, speech, October 7, 1962, Harrodsburg, Howells Family Collection.

21. Crawford, "There Was Neither Straight Shooting nor Ricochet Romance." Chinn at this time was still the deputy director of KHS.

22. Ibid.

23. Crawford, e-mail to author, April 6, 2013.

24. Chinn, interview by Ellis, 14.

7. Action at the Kentucky Historical Society, 1959–1973

1. Anderson, "Colonel George Morgan Chinn."
2. KHS minutes, April 6, 1962, Kentucky Historical Society.
3. KHS minutes, October 22, 1970, Kentucky Historical Society.
4. Klotter, "Thoughts on Colonel Chinn" (statement to author).
5. Klotter, e-mail to author, June 25, 2013.
6. Pearce, "Col. Jack Chinn."
7. Klotter, "Thoughts on Colonel Chinn" (statement to author).
8. "A History of the Kentucky Historical Society."
9. Day, "Mercer County."
10. Chinn, interview by Ellis.
11. Meeker, "Kentucky, Pennsylvania."
12. Ibid.
13. Ibid.
14. *Lakeland (PA) Leader,* June 12, 1963; Chinn, interview by Ellis.
15. Norman, "The Kentucky Colonel."
16. Ibid.
17. "Yank Cavalry Pierces Gap."
18. Lugart, "Kentucky Wins Flintlock Shoot." As governor, Combs had the first shot. He later reported, "I shocked those Pennsylvanians so much I had them scared before they started." He was referring to the fact that his shot was a bull's-eye.
19. *Park City Daily News* (Bowling Green, KY), September 20, 1964.
20. KHS minutes, July 6, 1962, Kentucky Historical Society.
21. KHS minutes, April 17, 1964, Kentucky Historical Society.
22. KHS minutes, January 15, 1966, Kentucky Historical Society.
23. http://www.historyburgoo.com/historyburgoo/2011/08/exploring the-kj. . . .
24. KHS minutes, July 18, 1967, Kentucky Historical Society.
25. KHS minutes, April 16, 1966, Kentucky Historical Society.
26. KHS minutes, April 29, 1967, Kentucky Historical Society.
27. Chinn, interview by Ellis.
28. Ibid.
29. *Herald-Leader,* May 26, 1999.
30. KHS minutes, April 6, 1962, Kentucky Historical Society; telephone conversation between Buddy Howells (Chinn's grandson) and author, June 8, 2013.
31. KHS minutes, April 6, 1962, Kentucky Historical Society.
32. Hackett, "Boone's Rifle."
33. Bright, "Mercer County's Modest Marine."

34. "Boone Grave to Boonesborough State Park?"

35. Ibid.

36. KHS minutes, April 19, 1960, Kentucky Historical Society. Bayliss Hardin, a gifted illustrator, illustrated one of Chinn's books, *Encyclopedia of American Hand Arms*. See Martin, "Capt George Chinn's Improvement."

37. Clark, letter to Breckinridge, University of Kentucky Libraries.

38. KHS minutes, April 19, 1960, Kentucky Historical Society.

39. Ibid.

40. KHS minutes, October 12, 1973, Kentucky Historical Society.

41. Givens, e-mail to author.

42. Ibid.

43. In a little Kentucky town between Bowling Green and Louisville, there is a historical marker about Andrew Jackson. It reads: "One block west stands a log inn built on a pioneer trail. . . . Among the many distinguished guests was Gen. Andrew Jackson in 1829 enroute to his inauguration as seventh president of the United States." At the time in question, President-elect Jackson was a passenger on the Steamboat *Fairy* on the Ohio River heading for Louisville, where he changed to the *Pennsylvania* to take him on up the Ohio, all the way to Pittsburgh, and then over the mountains to Washington City. See Jackson, *Bittersweet Journey*.

44. West, e-mail to author.

45. *The Private War of Lizzie Hardin* won an Award of Merit from the Association of State and Local History. KHS minutes, October 27, 1962, Kentucky Historical Society.

46. KHS minutes, April 10, 1963, October 11, 1963, July 18, 1967, Kentucky Historical Society.

47. KHS minutes, October 11, 1963, Kentucky Historical Society.

48. Ibid.

49. Chinn, interview by Ellis.

50. KHS minutes, April 21, 1971, Kentucky Historical Society.

51. KHS minutes, July 1, 1970, Kentucky Historical Society.

52. Ibid.

53. KHS minutes, January 14, 1971, Kentucky Historical Society.

54. KHS minutes, January 21, 1972, Kentucky Historical Society. Nothing, of course, ever came of this rumor. It was generated simply by the heat of the day. In fact, KHS at this time was contemplating a new building for its headquarters in Frankfort.

55. Today, 2014, UPK is the "scholarly publisher for the Common-

wealth" for Bellarmine University, Berea College, Centre College of Kentucky, Eastern Kentucky University, The Filson Historical Society, Georgetown College, Kentucky Historical Society, Kentucky State University, Morehead State University, Murray State University, Northern Kentucky University, Transylvania University, University of Kentucky, University of Louisville, and Western Kentucky University. When the University of Kentucky Press began its transition to the University Press of Kentucky in the late 1960s into the early 1970s, the author remembers quite well the trepidations of scholars at colleges and universities other than the University of Kentucky itself. It was widely believed that UK would dominate the field of general research to the detriment of the comprehensive institutions throughout the state. Subsequent events have shown this idea to be unfounded. UPK has been constantly fair-minded in its editorial processes. The UPK requirement of the onetime affiliation fee, which all other members of the publishing consortium paid, was discussed and approved at the KHS board meeting on October 14, 1971.

56. Meadows, e-mail to author. The author thanks Dr. Meadows and the entire KHS staff for their help in obtaining important materials for this work.

57. KHS minutes, January 5, 1962, Kentucky Historical Society.

58. KHS minutes, January 30, 1965, Kentucky Historical Society.

59. KHS minutes, December 19, 1964, Kentucky Historical Society.

60. KHS minutes, January 30, 1965, Kentucky Historical Society.

61. KHS minutes, July 26, 1966, Kentucky Historical Society.

62. KHS minutes, July 8, 1967, Kentucky Historical Society.

63. KHS minutes, November 10, 1961, Kentucky Historical Society.

64. Klotter, "Thoughts on Colonel Chinn" (statement to author).

65. Ibid.

66. Ted Bassett, telephone conversation with author, October 24, 2012.

67. Brown, "The Colonel Stirred Controversy."

68. Clark, interview by Marshall.

69. Ibid.

8. The World According to Chinn

1. Adkinson, e-mail to author.

2. Chinn, "Recipe for Chinn's Beer Cheese."

3. Ibid. The author expresses his thanks to Kandie Adkinson of the Kentucky Department of State in Frankfort for sending these recipes to me.

4. Broadside, "United States Marine Corps, First Annual 'Talk Meet,'" July 11, 1980, Howells Family Collection.

5. Klotter, "Thoughts on Colonel Chinn" (statement to author).

6. Ibid.

7. Howells, interview by author, May 22, 2012.

8. Gabler, "George Chinn."

9. Raitz and O'Malley, *Kentucky's Landscapes*, 317–18. See also Jackson, *A Social History of the Scotch-Irish*, 91.

10. Jackson, *A Social History of the Scotch-Irish*, 91.

11. Chinn, interview by Ellis.

12. Day, "Mercer County."

13. "Saluting the Saga of Harrodsburg," 18.

14. Chinn, *The History of Harrodsburg*, xvi.

15. Ibid., 452, 1, xvi.

16. Day, "Mercer County."

17. Ibid.

18. Hughes, interview by author.

19. Combs, "Col Chinn Separates Historical Fact from Fiction."

20. Harrison and Klotter, *A New History of Kentucky*, 58–59.

21. Hughes, interview by author.

22. Chinn, interview by Ellis.

23. "Saluting the Saga of Harrodsburg," 18.

24. Pope, "At 80, Retired Colonel Refuses to Stay on the Shelf."

25. Ibid.

26. Day, "Mercer County." The author's father, Luther H. Jackson (b. 1888) was raised in a remote village of northern Alabama. Sometimes the residents of these villages would learn, say on a Sunday, that on the following Tuesday at a certain spot *a car would pass through.* The village's young people would gather up picnic food and drink and actually spend the day (if need be) at the place. Sometimes the car came, sometimes it didn't. When it did, the car's driver and his passengers frequently stopped to have conversations with the assembled youth.

27. Smith, e-mail to author.

28. *Harrodsburg Herald*, September 8, 1987.

29. Smith, e-mail to author.

30. *Harrodsburg Herald*, July 13, 2000.

31. Ibid.
32. Ibid.

Conclusion

1. Jordan, "State Capital Guide is Top Arms Authority."

Bibliography

Achenbach, Joel. "Gettysburg." *Washington Post*, April 28, 2013.

American Historical Review (June 1989).

Anderson, Linda. "Colonel George Morgan Chinn." *Kentucky Historical Quarterly* (April 1976).

Bailey, Jack. *Murders, Mischief, Mayhem, Madness, Misdemeanors, and Downright Meanness in Mercer County.* Vol. 1. Privately published, 2005.

"A Big Man Becomes Bigger." *Harrodsburg Herald*, November 20, 1967.

"Boone Grave to Boonesborough State Park?" *Louisville Courier-Journal*, December 19, 1977.

Boone, Nathan. *My Father, Daniel Boone: The Draper Interviews with Nathan Boone.* Lexington: University Press of Kentucky, 2012.

Bright, Sallie. "Mercer County's Modest Marine." *Kentucky Advocate*, January 15, 1984.

Brown, John Y. *The Legend of the Praying Colonels.* Louisville: Marvin Gray, 1970.

Brown, Meridith. *Frontiersman: Daniel Boone and the Making of America.* Baton Rouge: Louisiana State University Press, 2008.

Brown, Mike. "The Colonel Stirred Controversy—but Enjoyed It." *Louisville Courier-Journal*, December 16, 1973.

Catawba College Yearbook, 1931.

Chandler, A. B., and Vance Trimble. *Heroes, Plain Folks, and Skunks: The Life and Times of Happy Chandler.* Chicago: Bonus Books, 1989.

"Chinn Describes Prexy's Defense; Says That Dr. Montgomery Carried the Ball in Great Style." *Cento*, December 14, 1923.

Chinn, George M. "Goes Hunting." *Cento*, December 8, 1923.

———. *The History of Harrodsburg and the "Great Settlement" Area of Kentucky, 1771–1900.* Harrodsburg: Harrodsburg Historical Society, 1985.

———. Interview by Byron Crawford, 1986.

———. Interview by William Ellis Jr., July 10, 1987, Harrodsburg, KY.

———. *The Machine Gun.* Vol. 5, *Engineering Designs Development, 1952–1987.* Ann Arbor, MI: Edwards, 1987.

———. "Recipe for Chinn's Beer Cheese." In *Bicentennial Recipe Book.* Harrodsburg, 1976.

———. *Through Two Hundred Years—Pictorial History of Mercer County, Kentucky.* Harrodsburg: self-published, 1974.

———. Untitled poem. *Centre College Cento,* November 21, 1923.

Chinn, George M., et al. *The Machine Gun.* Vol. 2, *The Development of Automatic Weapons prior to and during World War II by the Soviet Union and Its Satellite Countries.* Washington, DC: U.S. Navy, 1950.

"Colonel Jack Chinn Seems to Be on the Warpath." *Winchester (KY) News,* February 6, 1909.

Combs, Louise. "Col Chinn Separates Historical Fact from Fiction." *State Journal* (Frankfort, KY), October 24, 1975.

Communicadet, 1920.

Crawford, Byron. "Col. Chinn's Life Was a Sure Bet for Entertainment." *Louisville Courier-Journal,* March 31, 2006.

———. "Henry Ran Ringers around Shelby." *Louisville Courier-Journal,* October 29, 1980.

———. "Kentucky's 'Buddha of Ballistics': A Man of Wit and Humor Too." *Louisville Courier-Journal,* n.d.

———. "There Was Neither Straight Shooting nor Ricochet Romance." *Louisville Courier-Journal,* n.d., 1979.

Daly, Arthur. "Center for Centre." *New York Times,* November 23, 1944.

"David Skunks Goliath at Harvard." *New York Times,* October 29, 1921.

Day, Teresa. "Mercer County: History Near a Secondary Religion in 200-Year-Old Kentucky Treasure." *Lexington (KY) Herald-Leader,* August 18, 1985.

Dedman, Bud. Interview by Gary West, n.d., Harrodsburg, KY.

Draper, Lyman. *The Life of Daniel Boone.* Mechanicsville, PA: Stackpole Books, 1998.

"Due to Tuberculosis." *Berea (KY) Citizen,* May 18, 1909.

Ellis, J. Tandy. "When Col. Jack Chinn Got 'Het Up.'" *Louisville Times,* April 3, 1942.

Ezell, Virginia H. [Title unknown.] *National Defense* (September 1995).

Faragher, John. *Daniel Boone: The Life and Legend of an American Pioneer.* New York: Holt, 1992.

Gabler, David. "George Chinn." *Danville (KY) Advocate-Messenger,* May 12, 1986.

Gaddis, John Lewis. *The Landscape of History: How Historians Map the Past.* New York: Oxford University Press, 2002.

Gaddis, Linda. Interview by Roxanne Grider, September 12, 1990.

Gleason, F. W. Review of *The Machine Gun: History, Evolution, and Development of Manual, Automatic, and Airborne Repeating Weapons,* vol. 1, by Lt. Col. George M. Chinn et al. *Military Library* (Winter 1952).

Glenn, John. *John Glenn: A Memoir.* New York: Bantam Books, 1998.

Hackett, George W. "Boone's Rifle Embarrassed Weapons Expert." *Paducah (KY) Sun,* September 29, 1983.

Hanna, William. Letter to the editor. *Lexington (KY) Herald-Leader,* April 21, 2012.

Harrison, Lowell, and James Klotter. *A New History of Kentucky.* Lexington: University Press of Kentucky, 1997.

Hartley, Frank. "Has Guns, Still Traveling . . . That's Historian George Chinn." *Louisville Courier-Journal,* June 8, 1983.

Harvard Alumni Bulletin: Official Publication of the Harvard Alumni Association and of the Associated Harvard Clubs, November 3, 1921.

Hatter, Russell. Cave diary, January 1966–November 1968.

Hicks, Jack. "Before You Twist Cap, Salute John G. Carlisle." *Lexington (KY) Herald-Leader,* July 22, 1983.

———. "Marijuana's Roots Stretch Far Back into Kentucky's Past." *Cincinnati Enquirer,* August 19, 1983.

Hill, John. "Kentucky Politics at Its 'Damnedest': The 1955 Gubernatorial Election." Master's thesis, Western Kentucky University, 1998.

"A History of the Kentucky Historical Society." *Register of the Kentucky Historical Society,* n.d.

Howells, Howard (Buddy). "'Buddy' Howells, Grandson of Col. George M. Chinn." Interview by George Kontis. *Small Arms Review* (November 2010).

———. Interview by *Small Arms Defense Journal* (Winter 2002).

"It's No Surprise That Kentucky Rates Third Growing Marijuana." *Burlington (VT) Free Press*, August 20, 1983.

Jackson, Carlton. *Bittersweet Journey: Andrew Jackson's 1829 Inaugural Trip*. Sikston, MO: Acclaim, 2011.

———. *P.S. I Love You: The Story of the Singing Hilltoppers*. Lexington: University Press of Kentucky, 2007.

———. *A Social History of the Scotch-Irish*. Lanham, MD: Rowman-Littlefield, 1995.

———. *Zane Grey*. Rev. ed. Boston: Twayne, 1989.

Jordan, Paul R. "State Capital Guide Is Top Arms Authority." *Louisville Courier-Journal*, October 22, 1958.

"Kicking Up the Dirt." *Kentucky Advocate*, October 27, 1961.

Kleber, John E., ed. *Thomas D. Clark of Kentucky: An Uncommon Life in the Commonwealth*. Lexington: University Press of Kentucky, 2003.

Lewis, R. K. Review of *Encyclopedia of American Hand Arms*, by George Morgan Chinn. *Army Ordnance* 22, no. 132, n.d.

Lofaro, Michael. *Daniel Boone: An American Life*. Lexington: University Press of Kentucky, 2012.

"Lt. Col. Chinn Is One of the Top Men in Ordnance." *Windsock* (Cherry Point, NC), August 12, 1955.

Lugart, Fred W. "Kentucky Wins Flintlock Shoot." *Louisville Courier-Journal*, October 13, 1973.

Martin, Lorenzo. "Capt George Chinn's Improvement on .50 Caliber Is Tested." *Louisville Times*, July 11, 1944.

———. "Gun Experts Hail Kentuckian's Aid." *Louisville Times*, July 11, 1944.

———. "Gun Gadget Wins Fame for Chinn." *Louisville Times*, July 11, 1944.

Meeker, Robert F. "Kentucky, Pennsylvania Square Off for Shootout over Historic Origin of Famed Rifle." *Sunday Oregonian*, September 29, 1963.

Miller, Jim. "Historically Speaking." *Harrodsburg Herald*, August 11, 1994.

Morgan, Robert. *Boone: A Biography*. Chapel Hill, NC: Algonquin Books, 2007.

"News and Roundabout." *Franklin Weekly News* (Frankfort, KY), May 2, 1908.

Norman, Phil. "The Kentucky Colonel of Pen and Sword." *Louisville Courier-Journal Magazine*, August 4, 1985.

Okrent, Daniel. *Last Call: The Rise and Fall of Prohibition*. New York: Scribner, 2010.

Pearce, Bette. "Col. Jack Chinn Not Only Bred the First Kentucky Derby Winner but Also Authored the Law Creating Kentucky Racing Commission." *Lexington (KY) Herald-Leader*, February 4, 1973.

Pearce, John Ed. *The Colonel: The Captivating Biography of the Dynamic Founder of a Fast-Food Empire*. Garden City, NY: Doubleday, 1982.

Pioneer (Catawba College newspaper), October 20, 1930.

Pope, Lisa. "At 80, Retired Colonel Refuses to Stay on the Shelf." *Lexington (KY) Herald-Leader*, December 12, 1982.

"Prisoners Brutally Punished." *Newark (OH) Advocate*, April 25, 1908.

Pyle, Charles J. "Modernizing Our Small Arms: MK 19 40 MM Machine Gun." *Marine Corps Gazette* (April 1981).

Raitz, Karl, and Nancy O'Malley. *Kentucky's Landscapes along the Maysville Road*. Lexington: University Press of Kentucky, 2012.

Ramsey, Sy. "Ex-Marine Heads Historical Society." *Louisville Courier-Journal*, April 16, 1963.

Roberts, Mary. "Chinn's Inn." *Louisville Courier-Journal Magazine*, January 14, 1962.

Robertson, Robert W., Jr., *The Wonder Team: The Story of the Centre College Praying Colonels and Their Rise to the Top of the Football World, 1917–1924*. Louisville: Butler Books, 2008.

"Rock Thrower . . . Who Cut George Chinn's Face Held to September Court." *Harrodsburg (KY) Advocate-Messenger*, May 25, 1923.

"Saluting the Saga of Harrodsburg." *Southern Living* (June 1974).

Samuels, Benjamin. "Remembering a Forgotten Upset." *Harvard Crimson*, October 28, 2011.

"Teams in Good Condition: Centre and Texas Aggies Ready for Gridiron Test at Dallas." *New York Times*, January 1, 1922.

Thierman, Sue McClelland. "Modern Quarrying in the Ancient Way." *Louisville Courier-Journal Magazine*, November 16, 1958.

Thurman, Tom. *Run That by Me Again*. Kentucky Educational Television, 1977.

"Two Wounded in Shooting Affair near Brooklyn Bridge on Monday." *Harrodsburg Democrat*, May 13, 1930.

"Yank Cavalry Pierces Gap, Begins Drive on Frankfort." *Louisville Courier-Journal,* May 31, 1963.

Ziegler, Valarie H. "C6-H0: The Centre-Harvard Game of 1921." http://www.centre.edu/web/librfary/sc/special/c6ho/ziegler.html.

Zuercher, Rick. "That Day Frankfort Turned Dodge City." *State Journal* (Frankfort, KY), September 13, 1981.

Archival Sources

Danville Public Library, Danville, KY

Boyle County Collections.

Howells Family Collection, Harrodsburg, KY

Bergin, C. K. Letter to George M. Chinn, December 16, 1955.

Brown, William, rear admiral. Letter to George M. Chinn, May 7, 1946.

Chilton, William D., executive secretary of Kentucky Employees Retirement System. Letter to Chinn, February 20, 1961, file 400-54-1470.

Chinn, George M. Marine birthday speech, n.d., Harrodsburg.

———. Notes for Mother's Day speech, May 30, 1956, Harrodsburg.

———. Speech, October 7, 1962, Harrodsburg.

Commencement program, June 4, 1920, Harrodsburg.

Forrestal, James. Letter to George M. Chinn, n.d.

Geiger, Lieutenant, general of headquarters, Fleet Marine Force. Letter, n.d.

Hoover, J. Edgar. Letter to Admiral M. F. Schoeffel, July 16, 1954.

Matter, A. R. Letter to commandant, Marine Corps, July 1, 1945.

"United States Marine Corps, First Annual 'Talk Meet'" (broadside), July 11, 1980.

Wood, Frank E., Jr. Letter to George M. Chinn, February 7, 1945.

Kentucky Historical Society, Frankfort

KHS minutes, January 5, 1962, October 27, 1962, October 11, 1963, December 19, 1964, July 18, 1967, July 1, 1970, January 14, 1971, April 21, 1971, October 17, 1971, January 21, 1972.

Oral History Project

Chinn, George M. Interview by Guy Newland Jr., June 16, 1977.

Chinn, Haldon. Interview by Jane Hampton, February 7, 1990.

Flowers, Edna. Interview by Betsy Brenson, January 24, 2002.

Hinds, Charles. Interview by Betsy Brenson, January 8, 2002.
Sower, Frank. Interview by Betsy Brenson, January 4, 2002.

Louisville Public Library, Louisville, KY

Chinn, George Morgan. Letter to G. Glenn Clift, September 23, 1960.
Manning, Charles. Notes on Colonel Chinn's reaction to the "Communique" matter.

Marine Corps Museum, Quantico, VA

Chapman, General L. F. Letter to George Morgan Chinn, March 9, 1968.
Chinn, George Morgan. Letter to General L. F. Chapman, February 23, 1968.
Ignatius, Paul R. Letter to George M. Chinn, October 11, 1967.
Magruder, John H. Letter regarding George M. Chinn, November 2, 1958.
———. Letter to George M. Chinn, April 28, 1961.

Mercer County, Harrodsburg, KY

Collections on George Morgan Chinn.

University of Kentucky Libraries, Special Collections, Lexington

Chandler, A. B. Letter to George M. Chinn, December 3, 1931.
Chinn, George Morgan. Letters to A. B. Chandler, November 17, 1931, November 27, 1931, November 10, 1937, November 20, 1937.
Clark, Dr. Thomas D. Interview by William Marshall, n.d.
———. Letter to John B. Breckinridge, June 1, 1962.

Author's Interviews and Personal Correspondence

Adkinson, Kandie. E-mail to author, February 17, 2012.
Anderson, Lee. E-mails to author, April 8, 2013, November 5, 2013.
———. Interview by author, April 8, 2013, Bowling Green, KY.
Anonymous. E-mail to author, December 23, 2013.
Bassett, Ted. Telephone conversation with author, October 24, 2013.
Blankenship, Susan. Interview by author, August 28, 2013, Wilmore, KY.
Bryant, Ron. E-mails to author, February 13, 2012, March 9, 2013.

Crawford, Byron. E-mails to author, December 8, 2011, February 8, 2012, April 6, 2013.

Givens, Roger (Morgantown, KY). E-mail to author, June 22, 2013.

Hanna, William. Letter to author, March 2012.

Harrington, Drew. E-mail to author, February 20, 2013.

Harris, Russell. E-mails to author, October 8, 2012, May 7, 2013.

Hatter, Russell. E-mail to author, May 29, 2013.

Howells, Ann. E-mail to author, September 3, 2012.

Howells, Howard (Buddy). Interviews by author, February 7, 2012, February 12, 2012, April 11, 2012, May 21–22, 2012, September 12, 2012, September 16, 2012, September 17, 2012, September 18, 2012, September 19, 2012, September 25, 2012, February 22, 2013, July 22, 2013, Harrodsburg, KY.

Hughes, Nicky. Interview by author, October 3, 2012, Frankfort, KY.

Klotter, James. Interview by author, April 20, 2013, Bowling Green, KY.

———. "Thoughts on Colonel Chinn." Statement to author, n.d.

Kontis, George. E-mail to author, May 15, 2013.

McKee, John. E-mail to author, February 11, 2013.

Meadows, Darrell. E-mail to author, July 3, 2013.

Norrell, Robert. E-mail to author, January 20, 2013.

Pratt, Bill. E-mail to author, November 5, 2013.

Samuels, Benjamin. E-mail to author, October 13, 2012.

Smith, Ruth. E-mail to author, August 7, 2013.

Stringer, Donald. E-mail to author, November 4, 2013.

Weigel, Richard. E-mail to author, May 8, 2013.

West, Gary. E-mail to author, June 27, 2013.

Index

Index

machine guns *(cont.)*
  on Chinn house, 54–55;
  twenty-millimeter turrets,
  70. *See also* firearms; military
  weapons development
Madison (slave), 12–13
Magruder, John H., III, 85, 94
Manifest Destiny, 137
Mann, Vear, 119
Manning, Charles, 89
marble, 13
marijuana, 62–63
Marine Corps service. *See* U.S.
  Marine Corps service
marital advice, 80–81
Marsh, John, 143
Martin, George, 6
Massy, Rex, 112
Matter, A. R., 102
M-19 automatic grenade
  launchers, 70–74
McGinnis, Charlie, 143
McMillin, Alvin Nugent "Bo,"
  21, 23, 24
Mercer County (KY), 5
Mercer County Humane
  Society, 131
military museums, 85, 102–3
military service, Chinn's:
  assignments during World
  War II, 46, 56, 63–66;
  honors and awards, 1, 101–
  2, 140, 155–56; involvement
  with military museums, 85;
  during Korean War, 69–70;
  near-deployment in World
  War I, 4; post–Korean War
  retirement, 78; return to
  active duty during Vietnam
  era, 94–95; return to active

duty during World War II,
  57–60. *See also* U.S. Marine
  Corps service
military weapons development:
  Chinn's affinity and aptitude
  for, 1–2, 3, 63, 76–78,
  103–4; Browning .50 caliber
  machine guns, 56, 63–64;
  flameout eliminators,
  74–76; Chinn's inventions
  and patents, 156, 169n26;
  M-19 automatic grenade
  launchers, 70–74; Chinn's
  weapons consulting work,
  46, 56, 65; praise for Chinn's
  work, 66–67, 76; twenty-
  millimeter turrets, 70
Miller, Jim, 12
Millersburg Military Institute,
  3–4, 160n12
Montgomery, R. Ames, 33
Moran, Charles "Uncle
  Charlie": at Bucknell, 29,
  34; at Catawba College,
  34–36, 38; at Centre
  College, 20–21, 27, 28–29,
  31, 33
Morgan, Hannah, 10
Morgan, Ruth Grace, 10
Morrow, Edwin P., 24
mules, 122
Mundy's Landing, 2, 47
Murphy, Tim, 26
Myers, Robert, 25

"needle beer," 43
Nichols, Charles, 53
Nickell, Clarence E., 49
Niles, John Jacob, 115
Noland, William, 142

192